PRAISE

"The brilliantly paced chapters allow you to progressively absorb the facets, details and data through consistent building blocks, until you gradually grasp the vision of the book. I like how the chapters are arranged and spread into easily digestible subtopics. I appreciated the vision and synthesis. The author shows how to resolve historically disparate business processes and achieve new efficiencies through inter-related digital transformation strategies. I could easily go back and find certain topics I want to reread. Also, at the end of each chapter, Kevin L. Jackson shares a nice, inspirational summary, along with anecdotes, histories, and analyses. That adds a wonderful personal touch to a book that smoothly integrates a lot of business concepts. *Click to Transform* is a must-read for all business owners."

William Bierce | Award-winning technology and business attorney and bestselling author of *Smarter Business Exits*

"The timing is great to release this informative and relevant book, *Click to Transform*, written by Kevin L. Jackson, which showcases in great breadth the marvelous advantages of going digital. We are in the early stages of the digital transformation and Kevin shows us how businesses can significantly improve how they live, invest and grow if they lean forward and embrace these cloud-based business performance measures. Easy to understand, interesting to read, all while including key data and supporting information. Great book, highly recommended."

David Stehlin | Chief Executive Officer at TIA (Telecommunications Industry Association)

"*Click to Transform* discusses the "cloud" so comprehensively— its untapped advantages as well as the direction this astonishing cloud technology is heading and where businesses, companies, schools, hospitals, etc., fit into such a future. Read this now and see where your company should be heading!"

Melvin Greer | Chief Data Scientist, Americas at Intel Corporation

"Whereas most books about digital transformation tire me, Click to Transform fascinates me with its grand vision of the modern world and where the future of business is heading. Here Kevin L. Jackson was able to use both his wide and deep experience on the subject of hybrid infrastructure and present it in such a fresh, gripping way. Read this one, and it will do your company wonders."

Tom Fedro | Co-Founder, Paragon Software Group Corp.; bestselling author of *Next Level Selling*

"Kevin L. Jackson sure knows his stuff and was able to explain them easily. Reading Click to Transform, I was floored with the immensity of its scope and how superbly well it was written and presented. The author seems to have made sure not to miss anything of value, and I felt that in this work. Impressive, very relevant read, and highly recommended to all those running a business."

Sanjay Jaybhay | Bestselling author of *Invest and Grow Rich*

"Kevin L. Jackson has covered all critical corners concerning the digital transformation of your business. It answered all my questions and more. Click to Transform is not only very informative but also compelling in that it urged me to quickly adapt to the changing modern times. Do your company a huge favor by reading this magnificent book."

Ali Razi | Founder & CEO, Banc Certified Merchant Services

"*Click to Transform* is one of those books that can push you to change your business. Just like most people, I fear changing my ways with how I work and do business, but that's just another way of saying that I fear growing and transforming to what I and my company can truly be. This book has been a great guide for me."

Rick Orford | Co-Founder & Executive Producer at Travel Addicts Life, and bestselling author of *The Financially Independent Millennial*

"This book was able to show how the world moves digitally and what amazing modern technologies are already in place for businesses to transform into the New Digital Age. As a business owner reading this, I felt the need to catch up before it's all too late."

Mar Ricketts | Principal & Founder of GuildWorks

"*Click to Transform* did not only transform my company but also myself as a person. Everything just *clicked* in me when I read this book. I realized that this is not only about the Modern Digital Era but also about the Modern Digital Man. It's stunning what humans can do now compared to the past fifty years. The rate at which the world is changing is unprecedented, and each business must adapt in the best way possible in order to survive and flourish. *Click to Transform* helps us do just that."

Mark Nureddine | CEO of Bull Outdoor Products, and bestselling author of *Pocket Mentor*

Click to Transform

Digital Transformation Game Plan for Your Business

Kevin L. Jackson

Leaders Press

Leaders Press

Copyright © 2020 Kevin L. Jackson
All rights reserved. Published in the United States by Leaders Press.
www.leaderspress.com

All rights reserved. No part of this book may be reproduced or transmitted in any form or by any means, electronic or mechanical, including photocopying, recording, or by an information storage and retrieval system – except by a reviewer who may quote brief passages in a review to be printed in a magazine or newspaper – without permission in writing from the publisher.

ISBN 978-1-943386-90-1 (pbk)
ISBN 978-1-943386-89-5 (ebook)

Library of Congress Control Number: 2020918362

To my wife, Lisa.
My lifelong partner, teacher,
and conscience.

GET YOUR FREE DOWNLOAD AT:

CLICK TO TRANSFORM

LEADING TRANSFORMATION

Kevin L. Jackson

www.leaderspress.com/leading-transformation

Contents

Chapter 1: Digital Transformation Vision 13

Chapter 2: Transformation Capabilities and Functions 41

Chapter 3: Transformation Business Models 69

Chapter 4: Transformation Infrastructure 91

Chapter 5: Transformation Network 117

Chapter 6: Transformation Innovation 139

Chapter 7: Transformation Frameworks 189

Chapter 8: Embrace Transformation 207

Acknowledgements 211

About the Author 213

1

Digital Transformation Vision

Introduction

Digital transformation of an organization into a hybrid information technology model integrates digital technology into an organization's business or mission. Its purpose is to create and deliver innovative and industry-changing digital products and services to a global customer base. It also provides a seamless two-way flow of data and information between internal business processes and external processes that support customers, business partners, and the organization's business ecosystem.

Hybrid cloud computing supports this goal by delivering ubiquitous, convenient, and on-demand network access to shared and configurable computing resources. This approach provides access to digital products and services from anywhere, at any time, on any device. A strategy that merges hybrid cloud computing and digital transformation is a proven and viable solution for delivering quantum improvements in operational capability, market relevance, and business profit.

Digital transformation also requires enterprise adoption of an information technology environment. Such an environment blends cloud service provider environments with more traditional services originating from private enterprise datacenters and contractually engaged managed service providers. This model, referred to as hybrid information technology (hybrid IT), also drives the information technology (IT) management processes away from an enterprise-directed governance model toward one that accepts a shared governance and measured services philosophy. This modern philosophy embraces highly standardized architectures, IT service consumption, and automated infrastructure provisioning, configuration, operations, and de-provisioning. These typically massive technical and operational changes can also demand equally massive cultural shifts.

This book addresses both the integration of digital technology, the operational modifications, and the cultural changes necessary to deliver expected improvements, capabilities, relevance, and profits to organizations operating within nearly any industry.

Implementing cloud computing to achieve digital transformation can fundamentally enhance how an organization operates and delivers value to its customers and stakeholders. By doing this, an enterprise gains an ability to quickly act and react to changing data, operational conditions, and competitive strategies in a manner that supports the rapid attainment of the organization's goals. A survey of 2,000 executives conducted by Cognizant in 2016[1] identified the top five ways digital transformations generate value:

- Accelerating speed to market
- Strengthening competitive positioning
- Boosting revenue growth
- Raising employee productivity
- Expanding the ability to acquire, engage, and retain customers

Digital transformation is also a cultural change. The key requirement for success is challenging the status quo and actively encouraging experimentation. Culturally, "fail forward and fail often" must be viewed as an acceptable path toward success. Since no organization operates in a vacuum, digital transformation also delivers important and lasting effects on how an organization interacts with its partners and the broader ecosystem. In a digital world, those interactions, more often than not, take the form of software application capabilities and functions. Since cloud computing is now a widely used operating model for virtually all organizations in every industry, cloud computing for digital transformation is now an essential element to every organization's future.

This book supports those who have led, managed, or participated in a digital transformation effort. While there are many books on cloud computing in

---

[1] https://www.cognizant.com/futureofwork/article/the-end-goal-of-digital-transformation

the marketplace that address the technical aspects of cloud computing, virtually none of them discuss how to deliver real business or mission outcomes by using cloud computing models. There are at least three separate and distinct models that must be mastered if an organization wants to attain all of the much-heralded advantages of digital transformation to a hybrid cloud computing model:

1. *Economic model.* It delivers value through increasing use of the operational expenditure, or OPEX, financial model. This transition minimizes the amount of financial capital needed to affect a given business model while simultaneously enabling customer service delivery models that guarantee a determinate amount of income for a specified operational expense.
2. *Operational model.* It uses scalability, agility, flexibility, and availability as competitive advantages through the inherent design of these capabilities into the business or mission model.
3. *Technology service consumption model.* It reduces the traditional technology usage model (a selection, acquisition, operation, and maintenance of information technologies process) and increases the use of the technology services consumption model (evaluation, consumption, reevaluation, and selective replacement process).

As a cloud solution architect, business/mission leader, or digital transformation executive charged with leading a digital transformation, you will achieve professional success by evaluating multiple infrastructure options. Namely traditional datacenters, managed services providers (MSPs), or cloud service providers (CSPs), to meet your organization's goals. Collectively referred to as hybrid IT, all these resource options transform the customer experience, internal operational processes, and the targeted business models. Summarizing, this book shows you how to deliver digital transformation using cloud computing while simultaneously maintaining, or even increasing, the expected financial and operational return on other organizational IT infrastructure investments—effectively creating a hybrid cloud computing model.

This book also walks you through how to discuss, decide, coordinate, manage, and implement a digital transformation solution that leverages these three cloud computing models. It is also based on the assumption that organizations of any significant size need to design and deploy a hybrid IT solution to compete in their chosen industry vertical successfully. Based on the digital transformation triangle concept shown in figure 1.1, this book addresses how these components interact, affect IT governance, spur cultural management challenges, and advance an organization's digital transformation journey.

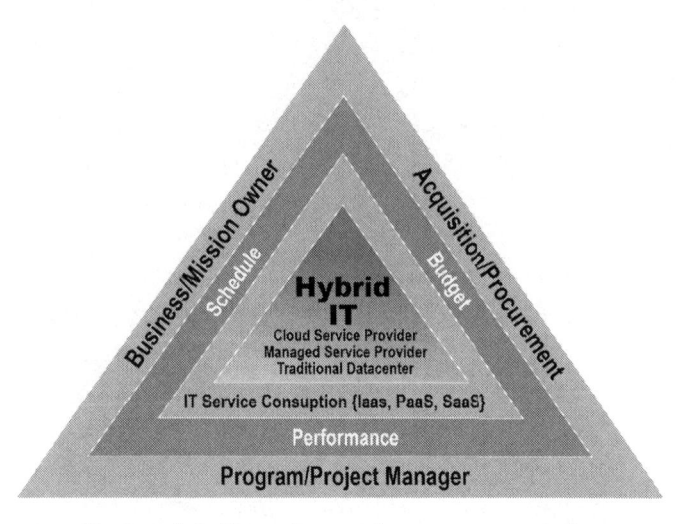

Figure 1.1. Digital transformation triangle.

Foundations of Hybrid Cloud Computing

Cloud computing, also known as information technology-as-a-service (ITaaS), was quickly adopted by many businesses and organizations because of its ability to deliver value in three significant market sectors:

- *Infrastructure-as-a-service* (IaaS). It enables a global scale and standardizes IT infrastructure services at an affordable price by using new, converged networks to deliver variable IT capacity pools.
- *Platform-as-a-service* (PaaS). It's a set of application development platforms that provides

multiple technology environments and greater flexibility to application developers

- *Software-as-a-service* (SaaS). It can reduce software consumption costs, especially in the area of application and software licensing, and also reduce software application support costs while simultaneously improving business back-end system capabilities.

These models, known collectively as cloud computing service models, deliver dramatic improvements in cost control, flexibility, speed to market, reliability, and resilience—all of which are vital cloud computing adoption drivers. These are all IT consumption models, not acquisition models. Customers pay for access to these services based on established consumption metrics. These metrics could be based on time (e.g., per hour for IaaS virtual machines), by the rate (e.g., input/output operation per second, or IOPs, for transactional services), by quantity (gigabytes, or GBs, stored for data storage), or any other metric set by the cloud service provider.

Consumption models differ from traditional IT acquisition models wherein a specific technology is purchased, owned, and operated by the customer. The difference is a critical posit because most organizational policies are built around IT acquisition. IT service consumption can significantly change corporate procurement and acquisition processes. The next sections cover each of these models.

Infrastructure-as-a-Service

In infrastructure-as-a-service, fundamental computing resources (i.e., compute, storage, and network services) are provided to a customer who can install and run any application onto a specified operating system. Management of the cloud infrastructure remains the responsibility of the service provider, but the customer has control over operating systems, storage, applications, and possibly limited control of select networking components.

Infrastructure traditionally has been the focal point for ensuring which capabilities and organization requirements must be met versus those that were restricted. It could also represent the most significant investments in terms of capital expenditures and skilled resources made by the organization. The emergence of cloud changes this traditional view of infrastructure's role with commoditization and enabling service consumption through an on-demand, pay-as-you-go model.

The IaaS service model depends on a large scale and the significantly high degree of automation needed to support significant internal user workloads or those across multiple cloud deployments. Customers require optimal levels of control, visibility, and assurances related to the infrastructure and their ability to satisfy their requirements. From a customer or user perspective, the compute and storage service pool appear seamless and endless. These services are driven and focused on supporting and meeting relevant service level agreements. High reliability and resilience are delivered through automated distribution across the virtualized infrastructure. For organizations, IaaS also provides usage metering and is priced based on units, or instances, consumed. This, in turn, can also be billed back to specific departments or functions. Prominent IaaS service providers include Microsoft, AT&T, Rackspace, and IBM.

Platform-as-a-Service

When consuming PaaS, the end customer is operating in an integrated development environment (IDE) provisioned and maintained by the cloud service provider. In this service model, customer-created or acquired applications can be deployed onto the cloud infrastructure. With these, the service provider provisions programming languages, libraries, services, and software development tools.

The cloud service consumer does not manage the underlying cloud infrastructure. That is managed and controlled by the CSP. They do, however, have control over the applications and possibly configuration settings for the hosting environment. Cloud platform components offered using the PaaS model have revolutionized software development and delivery over the past few years. This approach can drastically reduce costs and required resources. By accelerating innovation and time-to-market, the approach also promotes innovation. The PaaS model can support multiple languages and frameworks. This helps the developers code in whichever language they prefer or whatever the design requirements specify. Recently, significant strides and efforts have been taken to ensure that open-source stacks are both supported and utilized, thus reducing lock-in, which can be described as an inability to stop consuming services from a specific CSP due to economic, operational, or similar reasons. Open-source software stacks can also help avoid interoperability issues if the enterprise changes CSPs. The ability to support a wide range of underlying hosting environments for the platform is also a key PaaS benefit.

PaaS allows for choice and reduced lock-in. While a requirement to code to specific APIs was enforced by cloud service providers, developers were able to run applications in multiple environments. This ensured a level of consistency and quality for customers and users. Auto-scaling, the ability of an application infrastructure to scale up and down based on user demand, is also easily leveraged through PaaS. The platform allocates resources and assigns these to the application, as required. This capability effectively serves seasonal organizations that experience spikes and drops in usage. Notable PaaS providers include companies like Microsoft, Lightening, and Google.

Software-as-a-Service

Software-as-a-service is the largest by annual sales and the most accessible service model for consumption. This approach is the model employed when the consumer uses the provider's applications running on a cloud infrastructure. These applications are accessed using a browser, known as a thin client. In this model, the CSP retains control of the infrastructure and application capabilities. The customer will normally only have access to designated configuration settings.

SaaS organizations have potentially limitless possibilities for running programs and applications that may not have been previously practical or feasible. Clients can access applications, data, and cloud computing systems from any computer linked to the internet. Many SaaS consumers enjoy an overall reduction in the total cost of software ownership. This eliminates the need to purchase application software licenses or support. A pay-per-use licensing model replaces capital expenditures for software licenses. There is a limited administrative burden on the customer with updates and patch management being the service providers' responsibility. SaaS cloud service providers include Microsoft, Google, Salesforce, Oracle, and SAP.

X-as-a-Service[2] Operations

Offering or consuming cloud computing services both economically and profitably across a broad marketplace, however, involves much more than just technology changes. It also requires the build-out or adoption of a complementary wide-area information technology infrastructure and the widespread use of information exchange interface standards that we

---

[2] X-as-a-Services, or XaaS, is often used to reference any type of technology service consumption offering.

now refer to as the Internet. The commercialization of the internet resulted in the following:

- Development and near ubiquity of easy-to-use, visually attractive computing devices
- The rapid growth of globally interconnected wide-area networks
- Abandonment of then widely used, tightly coupled client-server business applications in favor of loosely coupled application architectures
- The disappearance of thick client software interfaces and the broad adoption of thin-client browsers

These foundational technical changes were blended with a standards-based and highly automated operational approach. The Internet Engineering Task Force (IETF) sets Internet standards, but standards may also be set in a de facto manner if a vendor-specific technical implementation is adopted by a large segment of the IT marketplace. An example of a de facto standard is the widely used MP3 audio format, which started as an alternative to WAV for Internet music distribution, then replaced it. In the end, these were the processes that created the economic revolution that cloud computing represents today.

IT Service Consumption Policies

As alluded to previously, cloud service consumption demands the implementation of updated organizational procurement and acquisition policies.

Previously sacrosanct end-user requirements are no longer central to the acquisition process. Since this approach represents the consumption of services and not the acquisition of products, all IT services are selected from existing cloud provider service catalogs. CSPs are generally not open to varying the technical specifications, pricing, service metrics, or operational service levels. Short-term (monthly) periodic billing is based on specified service consumption metrics, and the service level agreement (SLA) is central to all contractual obligations. Periodic (monthly) billing reconciliation is also prudent.

Operational and policy adaptations that explicitly recognize IT service consumption characteristics align the organization's internal processes with key cloud computing characteristics.

Key Cloud Computing Characteristics

Coined by the United States National Institute of Standards and Technology (NIST), there are five "essential characteristics of cloud computing." These characteristics succinctly describe the technical, operational, and economic aspects of this revolutionary approach to information technology:

- On-demand self-service
- Broad network access
- Resource pooling
- Rapid elasticity
- Measured service

On-Demand Self-Service

On-demand self-service enables the provisioning of cloud resources on demand, whenever and wherever they are required. From a security viewpoint, this has introduced challenges to governing the use and provisioning of cloud-based services, which often violate organizational policies. Financial office approval is not normally required for on-demand self-service provisioning. A service can be ordered and provisioned using a personal credit card.

Broad Network Access

Cloud computing is global in nature. It is designed as an always on and always available offering. This characteristic, referred to as broad network access, delivers continuous access convenience and access to any desired services, when needed, from anywhere. All a consumer needs are Internet access and relevant credentials. The rapid and broad adoption of mobile and smart devices by businesses has raised the importance of this characteristic. It has also, however, challenged organizational policies typically designed to maintain tight control on any device that processes organizational data. This will drive a need for corporate policy changes, mobile device management (MDM), and digital rights management (DRM) across most companies.

Resource Pooling

Resource pooling drives the core economics of cloud computing. This, the third essential characteristic, addresses the improved resource utilization rates cloud computing can deliver. Traditional data centers see resource utilization rates of between 10 and 20 percent. This is caused by a design philosophy that assures resource availability during peak-usage time periods. With cloud computing, resources are grouped in pools for use across all customer groups, which is referred to as multitenancy. Resources in this environment can automatically scale and adjust to the user's or client's needs, based on their workload or resource requirements. Cloud service providers dynamically reassign resources to multiple tenants from large resource pools. Resource provisioning across each client is prioritized based on the associated client service level agreement (SLA), which facilitates the appropriate resourcing for each client. An important, and often ignored, resource pooling characteristic enables the lower computing costs executives tout as the main reason for cloud adoption. These leaders often sabotage transformation initiatives by demanding a dedicated computing environment. Avoiding multitenancy and resource pooling violates the cloud computing economic model resulting in operational cost that generally exceeds that of a traditional datacenter.

Rapid Elasticity

Rapid elasticity, the next characteristic, allows the user to obtain additional resources, storage, and computational power as the user's need or workload requires. This trait is more often transparent to the user, with more resources seamlessly added. Because cloud services use the pay-per-use concept within a multitenant environment, in which consuming entities only pay for what is consumed. "Multitenant" is used to describe an IT operational environment that is accessible, configurable, and used by many unrelated operational entities. Pay-per-use is of particular benefit to seasonal or event-type businesses utilizing cloud services. For example, consider an online retailer preparing for holiday sales. Prior to the holiday, relatively few computing resources are needed. When the holiday arrives, however, the retailer's website may need to accommodate a large spike in user demand. This is where rapid elasticity demonstrates its advantages when compared with enterprises that use traditional IT deployments, which would have required heavy capital investment in advance of the demand in order to support such a spike.

Measured Service

The last essential characteristic is measured service. Cloud computing offers resource usage metering that can be measured, controlled, reported, and alerted. This characteristic delivers multiple benefits and overall transparency between the provider and the client. Similar to how electricity service or a mobile phone is metered, measuring these services helps customers maintain awareness of their costs. The customer pays for his use and has the ability to get an itemized bill or breakdown of usage. A key benefit exploited by many organizations is the ability to charge departments or business units for their use of services. This allows IT and finance to quantify exact usage and costs per department or by business function. This is very difficult to achieve in traditional IT environments. When digital transformation is the goal, this aspect transforms the IT unit from a cost center to a profit center. This liberating step can prevent expenditures on bespoke or unique IT services and empowers IT management by providing data that informs the executive decision process.

Cloud Computing Deployment Models

There are four generally accepted cloud computing deployment models:

- Private cloud refers to a proprietary environment (network or data center) owned and architected for use by a specified entity that uses a cloud computing approach to provide services behind a firewall.
- Public cloud is a publically available service, generally over the Internet, in which a customer can access cloud service provider resources, as a free service or offered on a pay-per-usage model.
- Multiple organizations share a community cloud environment under a joint governance structure. This model gives many of the public cloud benefits while providing heightened levels of privacy, security, and regulatory compliance.

- A hybrid cloud is formed by combining any of the other deployment models, typically public and private cloud.

The next sections discuss each of these deployment models.

Private Cloud

In a private cloud, the cloud infrastructure is only accessible to a single organization. The organization may, however, be comprised of different operational entities, like different business units. This type of cloud may be owned, managed, and operated by the organization, a third party, or some combination. It can exist on or off-premises.

A private cloud is normally managed by the organization itself, but outsourcing the general management to trusted third parties is also an option. This deployment model is normally for the exclusive use of the owning organization, its employees, and the designated business ecosystem. The private cloud is sometimes referred to as the internal or organizational cloud. When using a private cloud, the organization enjoys increased control over data, underlying systems, and applications. Another benefit is the retention of ownership privileges and governance controls. With ownership, the enterprise also gains assurance over data location.

Private clouds have become more popular among large organizations with legacy systems, customized environments, and when significant technology investments have already been made. When considering all the cloud deployment options, it may be more financially viable to utilize and incorporate previous IT investments within a private cloud environment than to discard or retire these assets.

Public Cloud

The public cloud deployment model is when the cloud infrastructure is provisioned for open use by the general public and usually under a multitenant model. A public cloud may be owned, managed, and operated by a business, academic, or government organization, or some combination of them. It also exists within cloud-service-provider-owned facilities.

Public cloud consumption is a proven cost saver. Cost savings occur because expenditures for acquiring hardware, application, and bandwidth costs are borne by the provider. Providers also streamline use by making it easy to provision resources. Public cloud providers typically have the scalability to meet customer needs, customers benefit financially because they pay only for what they consume. With increasing demands for public cloud services, many data center providers offer and remodel their services as public cloud offerings. Significant public CSPs include Amazon Web Services, Microsoft, Digital Ocean, and Google.

Community Cloud

In a community cloud, the cloud infrastructure provides services to a designated customer community that has agreed to abide by common IT governance standards. These organizations typically share common IT governance and policy concerns. These concerns may include mission, security requirements, data sovereignty, or industry compliance considerations. Community clouds may be owned, managed, and operated by one or more of the community members, a third party, or some combination. This deployment model may exist on or off-member-premises. Community clouds should give the benefits of a public cloud deployment while providing heightened levels of privacy, security, and regulatory compliance.

Hybrid Cloud

The final deployment model, hybrid cloud, is any combination of the three previously described models. Don't confuse "hybrid cloud" with "hybrid IT." The latter term refers to the use of cloud infrastructure in combination with management services or traditional data centers. This strategic option is separate from the selection of a cloud service or deployment model. Established organizations should also develop an overall IT implementation strategy. As a baseline, this implementation strategy should target an IT infrastructure mix best suited for the enterprise's long-term IT investment and staffing skill plans. Components of this broader IT implementation mix are the following:

- Traditional data centers, where the organization pays for the total cost of ownership of all the hardware and software. It also pays the salaries and benefits for the operations staff. The organization has complete and total control of information technology governance in this model.
- Managed service provider (MSP) in which the organization pays a third party to provide IT hardware, software, and professional services. In this model, the organization retains IT governance control through the negotiation and enforcement of a binding contract with the MSP. The enterprise pays all MSP costs plus a negotiated profit. This option can be cheaper than the organizationally owned data center if the MSP is more efficient than the contracting enterprise in delivering the contracted services.
- Cloud computing, where the cloud service provider (CSP) pays for ownership of all hardware and software, and the CSP pays the salaries and benefits for all operational staff. In this model, the CSP has total control of information technology governance.

IT Implementation Options

IT Implementation options (enterprise, MSP, or CSP) are dictated by IT governance priorities and IT infrastructure investment decisions. The organization must weigh the cost, benefits, and shortcomings of these three options for each of the corporate business processes or applications across the entire application portfolio. Application migration decisions should be driven by total ROI data and enterprise-wide business value analysis. This is the reason why some applications should migrate to the cloud, some should be hosted by an MSP, and others should remain in a traditional data center environment. Digital transformation does not mean that every application must transition to the cloud. It should, however, initiate a review of the entire enterprise application portfolio so that informed decisions can be made.

An overall IT implementation strategy mix blends the right deployment model, service model, and implementation option for each of your critical business processes. Deployment model selection is governed by the organization's risk tolerance. The basis for your service model selections (IaaS, PaaS, or SaaS) should be employee skill sets and supporting training budgets. Given this, organizations of any significant size normally need to pursue a hybrid information technology (hybrid IT) high-level strategy.

Platform Economics

The fundamental goal of all business decisions is to improve the economic well-being of an organization.

This is why a key concept you should understand when integrating digital technology into an organization is platform economics. The platform in this concept is a digital environment characterized by near-zero marginal cost of access, reproduction, and distribution. This form of business model is disruptive to traditional goods and services because of the zero cost and instant economics of digital information goods in today's pervasive digital network environment. Since most traditional goods and services are not free, perfect, or instant, companies that only sell traditional goods are placed at a competitive disadvantage. Networked goods become more valuable as more people use them. This results in "demand-side economies of scale," which gives an advantage to bigger networks. When combined with Moore's law and combinatorial innovation, platform economics is quite literally reshaping all areas of industry.

Platform economics delivers new revenue streams by inventing new customer value propositions. It also creates new industry value chains by generating new customer needs and segments. The digital nature of platform economics can be leveraged to design, develop, and deliver new products, services, and capabilities as well. Normally, this is accomplished through product and service optimization that enhances the customer's value proposition and improves current value chains by exploiting the efficiency offered by cloud computing. Organizations that successfully embed platform economics into their digitally transformed business models and processes can establish market differentiation through innovation. Innovation is key to extending the customer value proposition and improving the organization's value to each individual customer. All these platform economics characteristics result in new revenue streams and industry ecosystem role enhancements.

Platform Business Models

Business models that take advantage of platform economics include these:

- Freemium offers customers a basic level of service for free, then charges for extra function or capacity on top of the base (e.g., Dropbox, Evernote).
- The ad revenue model makes money by showing ads to users (e.g., Google iPhone search engine, Waze, Facebook on iPhone).
- An enhanced customer service model differentiates your products and services from those of your competitors (e.g., Chase, Amazon).
- Public service offerings freely deliver a valuable service to the general public in exchange for goodwill-enhanced corporate reputation and noncommercial marketplace interaction. A few examples are these: Street Bump uses iPhone sensors to report pothole location to the city, Federal Reserve Economic Data, or FRED[3] from the Federal Reserve Bank of St.

---

[3] https://fred.stlouisfed.org/

Louis that provides a wide range of economic data freely to the public, and Code for America that lets people take a sabbatical from technology firms to develop apps for city governments.
- The pairing of traditional products with digital extensions not only delivers differentiation but also efficiently builds social communities. This model can also leverage crowdsourcing techniques to improve current products and innovate to create new products and services. A few examples are health and fitness devices like Nike+ FuelBand and Fitbit, August smart door locks, Sonos music speakers, Drop kitchen weighing scales, gear controlled by apps like Viper.

By executing a platform economics business model on a cloud computing platform and organizations can capture revenue before consuming any IT services. This ensures that every financial expenditure on digital products or service delivery has a guaranteed revenue return. This approach reduces or eliminates costs associated with physical products and services. Operationally, companies have a greater breadth to pivot and shift by drastically reducing the reaction time needed to respond to environmental changes. By enhancing marketplace relevance, the organization's customer base naturally expands, margins (e.g., operational, financial, temporal, quality) are enhanced, and innovation becomes second nature.

Summary

Platform economics is a valuable disruption—much like cloud computing itself. When reflecting upon the history of cloud computing and how it fits into the future, I am reminded of the world's greatest disruptors. The history of disruption through standardization traces back to Henry Ford.

Henry Ford did not invent the automobile. When Ford started Ford Motor Company, smaller manufacturers were already making their own cars. The cars took consumers from point A to point B, but a big problem prevented the industry from moving forward. Each car was made with parts unique to its manufacturer. Fixing these parts required knowledge unavailable to local mechanics in different parts of the country. Replacing these parts took ages. Progress moved slower than a horse and carriage.

This industry reflects the early days of IT. Individual programmers built solutions, but they contained unique bits and bytes not universally understood or easy to replace. Unless you had a hand in working with specific software, you were left in the dark. Collaboration and standardization were nonexistent.

Henry Ford did not invent any parts of the automobile. He pooled the existing knowledge of manufacturers and standardized the way that pieces and parts built cars. Pieces were no longer made specifically for one vehicle. Replacements were easier to access and cheaper to obtain. Henry Ford did not invent the assembly line, but he made the assembly line possible. The assembly process could only exist if everyone's job and materials were standardized.

It was not Henry Ford's extensive knowledge of automobiles that changed the industry. It was his determination to collaborate, provide access, and set standards that transformed the way that automobiles were made. Ford needed to step back and see the problems slowing down the industry. Only then could he make things easier for everyone involved. Building a car is not about its parts—it's about how effectively you can put these parts together.

This idea continues to disrupt, standardize, and revolutionize common business practices. Andrew Jassey did not "invent" the API. But in the first years of AWS, development platforms were much like the parts of early automobiles. Developers worked separately with "parts" that didn't fit on other platforms. It was time for collaboration and standardization. Building websites for e-commerce was not about building a new technology—it was about how effectively you can standardize technology and make it accessible to merchants who used it to sell products.

Today, if AWS were to stand alone, it would be worth USD300 to USD400 billion.

Jassey and Ford could not have disrupted entire industries if they didn't see the industry (e.g., the problem) itself as a whole. Jassey understood the pains of e-commerce businesses and developers who attempted to build their sites. Technology was slowing businesses down. Ford understood the pains that came from building, fixing, and repairing cars with unique parts. Technology was slowing car manufacturers down. By standardizing the technology, Jassey and Ford made the origin of that technology irrelevant. Once more, focus on the individual project components must now be shifted toward the harmonization of all system parts working together, in unison.

This step away from technology is often the missing puzzle piece of transformational innovation. And therein lies the value of hybrid IT strategies using cloud computing. The existence of cloud technology is irrelevant. It's merely a tool in your toolbox. Great builders do not spend their lives looking for the shiniest hammer. They use that hammer, put it to a nail, and build skyscrapers. You have all of the tools available to you in the cloud. What are you going to build?

2

Transformation Capabilities and Functions

A successful digital transformation initiative results in fundamental changes to how the organization operates and how the technology enables the organization to meet its goals. Clear visibility and insights into the organization's operational processes and the needs of the ecosystem members inform these changes. This visibility and insight are worthless if not tied to action.

One of the most prevalent problems companies have when engaged in digital transformation is an inability to act or react fast enough to matter.[4] Digital transformation combines the information logistics systems capable of collecting, analyzing, and reporting data fast enough to be useful, along with the ability to act and react in response. This chapter outlines

---

[4] https://www.cognizant.com/futureofwork/article/the-end-goal-of-digital-transformation

the significant digital transformation business capabilities and functions that enable this critical action component along with the challenges they address and their related functional architectures.

Digital transformation business capabilities and functions reviewed in this chapter are the following:

- API management
- Big data and analytics
- Blockchain
- E-commerce
- Enterprise social collaboration
- Internet-of-things (IoT)
- Enterprise mobility
- Secure data management

An important note is that a "function" is what something does or is used to complete a task. Functions are used to enable capabilities. A "capability," on the other hand, empowers the organization with an ability to generate an outcome. This is important because an outcome can be delivered as a product or service to an end customer.

API Management Function

Modern organizations interact with their customers, supply chains, and partners through software. These applications represent every critical business or mission process. Applications interact with each other through the use of application programming interfaces, or APIs. This trait makes API management a digital transformation foundational capability.

APIs represent the public persona of an enterprise. They expose defined assets, data, or services for ecosystem consumption. Application developers invoke APIs via a web browser, mobile application, or device to exchange services and/or communicate through the documented interface, as figure 2.1 shows.

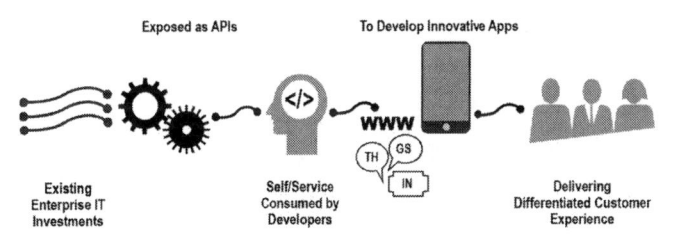

Figure 2.1. APIs enable innovation by exposing organizational assets and functionality.

API management platforms enable innovation by making it easier to expose existing organizational assets and functionality. Exposing existing functions, like analysis or search, through APIs, can create new products or enhance existing products and services. The resulting self-service portals can be used by any application developer. This transformative activity can open existing enterprise assets to new channels and enrich customer experience in integrated omni-channel interactions. An API management platform enables controlled and secure self-service access to core business assets.

Before initiating their pursuit of digital transformation, enterprises should expose enterprise functionality as a set of reusable APIs. This accelerates in-house application development and opens the way for self-service consumption. Digital applications should also be migrated to a cloud computing platform to enable quick deployment and engagement across new channels. Mobile and IoT device access should also be added.

Business ecosystem expansion and enablement are also driven by APIs. These recommendations are all proven pathways toward the monetization of processing algorithms and data.

When adopting an API strategy, enterprises use a comprehensive API management platform to create and test functionality and connecting implementation code to backend systems. It is also perfect for managing them during the production phase. Such a platform includes tools to design, model, develop, test, and deploy APIs in an automated, continuous delivery model. Polyglot runtime support is also key to enabling innovation and agility within different programming models. Strong nonfunctional characteristics like monitoring, scalability, load balancing, lifecycle governance, and failover should all be considered. Access management across multiple user designations and levels is also a key security nonfunctional attribute.

API management platform components (figure 2.2) interact with one another to support various use cases that involve application developers, API developers, API owners, and IT operations.

Cloud environments are flexible and allow for less concern regarding physical connectivity. The need for advanced planning remains essential, however. Criteria that should be reviewed to enable better provisioning of data and computing resources include the following:

- Elasticity
- Compute services
- Resilience
- Security
- Optimized service provisioning
- Optimized supply chain

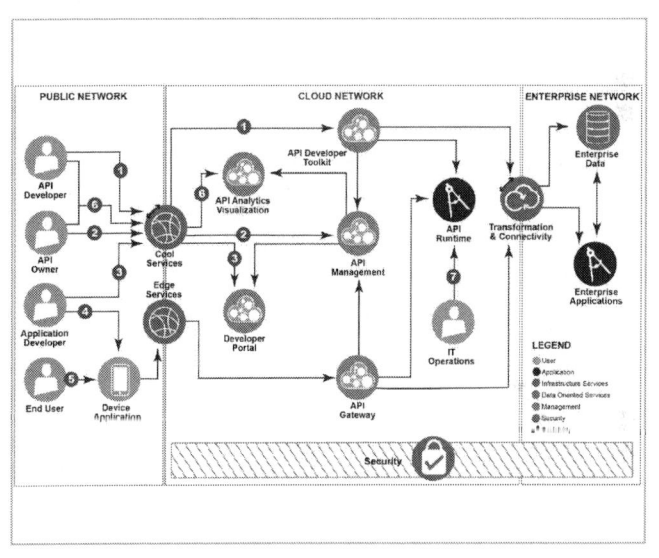

Figure 2.2. API management platform components.

Elasticity

Infrastructure elasticity refers to a cloud solution's ability to provision and deprovision computing resources on demand. Public clouds have a clear advantage in this capability since they generally have larger pools of resources available. Customers benefit through on-demand pricing within which they only pay for the services used. An API platform must provide scalable processing capabilities for fast performance. This supports enterprise expansion and fluctuating quantities of API calls.

Compute Services

Inexpensive commodity processors support modern development environments that use Hadoop, Spark, and Jupyter (iPython), by allowing them the freedom to take advantage of these massively parallel hardware systems. High-speed analytics and streaming data are two areas where cloud-based processor pools enable real-time, in-motion data solutions. Dedicated hardware delivers faster development and testing before migrating to hybrid and public environments.

A flexible API platform would require the deployment of multiple environments to support development lifecycle requirements and data governance restrictions. The cloud computing infrastructure model provides this provisioning flexibility.

Resilience

API management platforms should not be open to any single point of failure. This point drives resilience and fault tolerance requirements. CSP components are inherently resilient due to the widespread use of clustering and standard provisioning of multiple program instances and cloud services. This operational norm combines with data replication and redundancy across multiple storage locations, both physical and virtual. Physical and virtual networks should also have multiple paths and multiple providers.

Security

Data privacy and identity management are increasingly important to modern operations. As volume and variety expand, information governance and security challenges increase. This fact calls for better enterprise data monitoring and compliance strategies. In general, cloud computing allows for faster deployment of desired compliance and monitoring tools, which encourages agile policy and compliance governance frameworks. API security is tightly coupled to data access control. This is true because security must also function in as autonomous and automatic a fashion as possible, which is opposed to traditional network security. The speed and agility of modern cloud deployment are simply too fast and rapidly changing for traditional network security methods. As a result, API security policies must be more complex and granular than in the past. This is especially true when brokering authentication and authorization credentials across domains.

Hybrid IT and APIs

Enterprises routinely require a combination of public cloud, private cloud, and on-premises components. When linked together, such an environment creates a hybrid IT environment.

Businesses implementing hybrid IT and hybrid cloud solutions are looking for flexibility and agility in delivering new capabilities. This includes access by mobile workers, business ecosystem expansion, and market extension. APIs deliver both flexibility and agility.

Big Data Analytics Capability

Big data analytics is a huge priority for many companies. When cloud computing and big data technologies are merged, they offer a very cost-effective delivery model for big data storage, analysis, and, most of all, ubiquitous access to analytical results. That access can enable quick and executive decisions and effective action. Reasons for making big data analytics a key component of a digital transformation initiative include the following:

- The low up-front cost of delivery using cloud computing solutions, which allow for a quick establishment of an analytics infrastructure.

Organizations can also test new scenarios with little up-front expenses. This enables rapid and affordable iteration when exploring analytical results.
- Speed and agility in setting up analytics infrastructures. This contrasts with traditional delays caused by ordering, deploying, and operating physical infrastructure. Cloud agility also delivers more flexibility, agility, and variable infrastructure costs of an analytics infrastructure as data volumes fluctuate up and down.
- An improved ability to stay current with changing data platforms and analytics capabilities. In addition, as data volumes grow, emerging cloud technologies offer scale-out data transfer that also enables more efficient, large-scale workflows for ingesting and exchanging big data.
- Improved cloud security through the ability of cloud service providers to enforce security across multiple layers and processes as needed to demonstrate legal and regulatory compliance.
- Reduced new business model barriers to entry through the elimination of large initial investments.
- Accelerate innovation: with the ability to quickly leverage big data analytics through the low costs, agility, speed, and security that modern cloud solutions offer, companies have more resources available to experiment with other innovative technologies.

The options allow for flexibility in the placement of both data and required analytics workloads. Legal and regulatory requirements could also drive data location decisions.

This big data and analytics architecture (figure 2.3) in the cloud is very similar to traditional data lake deployments where both structured and unstructured data sources are staged and transformed by data integration and stream computing engines. Enterprise users can then access resources on premises or using a secure VPN. Data is made available directly and via reports and analytics applications. Transformation and connectivity gateways format data and information for use by enterprise applications, mobile devices, and desktop systems. Third-party users with proper credentials can also gain access to the shared data with edge services.

Big data and analytics use cases typically leverage a hybrid cloud, and the financial services industry uses this capability widely for mobile-banking applications. Key-supported–banking functions could include personalized interaction and call center agent Q&A solutions for credit card customer service or analytics that deliver cognitive insights to marketers to help understand and anticipate customer behaviors. Bank agents are also guided toward the most appropriate client communication strategies. These solutions can also provide the following services:

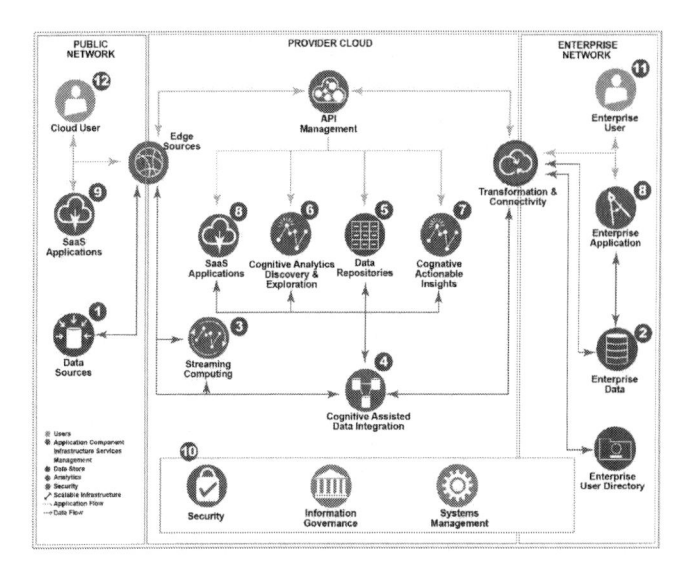

Figure 2.3. Big data analytics functional components.

- Deliver and maintain customer profile data. Enterprise database interactions are collected along with public social media data.
- Integrate and transform both enterprise and public data.
- Store both transformed data and interaction data in a data lake repository for later analysis.
- Service real-time API requests with data cached on an in-memory database.
- Perform rule-based analytics to analyze the personal client experience and present changes to personnel or mobile applications where appropriate to improve the client experience in real time.

- Help data scientists discover and explore patterns.
- The APIs read the data upon real-time requests that are typically from branches.
- Use a cloud-based marketing platform for omni-channel interaction with clients leads to better client experience.
- The cognitive APIs are used currently for Q&A for credit card marketing agents.

Blockchain Function

Blockchain features an immutable distributed ledger, which is a decentralized network that is cryptographically secured. It allows participants to engage in business transactions that span geographical boundaries through a shared ledger using peer-to-peer replication, which is updated every time a transaction block is agreed to be committed. This approach can reduce operational costs and friction, create immutable transaction records, and enable nearly instantaneous updates of transparent ledgers. When used for digital transformation, it provides an exchange network for securely transferring money, assets, or other valuables between willing parties. These transactions are immutable and do not require using a trusted third party. Cryptography ensures that network participants see only the relevant parts of the ledger and that transactions are secure, authenticated, and verifiable within the context of permission business blockchains.

Although cloud computing is not required to build blockchain platforms, services, or networks (figure 2.4), using the cloud is highly recommended because of its elasticity, performance, networking characteristics, and its distributed nature.

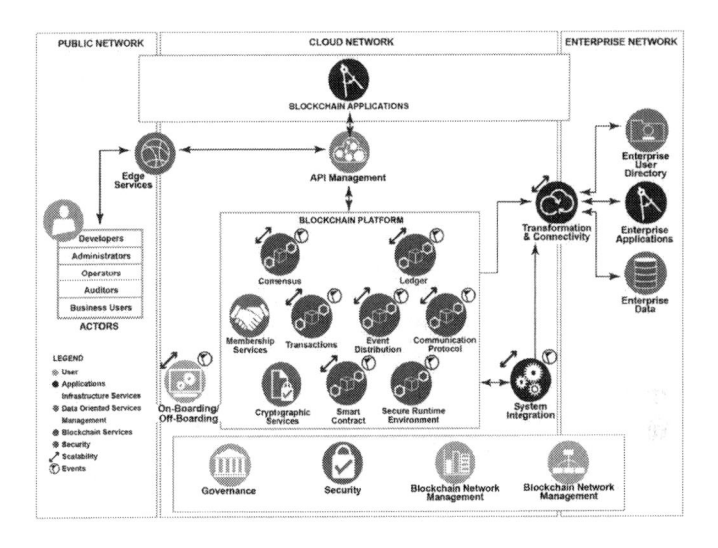

Figure 2.4. Blockchain platform functional components.

Blockchain technical implementations vary. Different methods are used to establish the network topology, manage participation, execute smart contracts, and manage growth. Businesses are likely to access the blockchain network through cloud-based blockchain applications. One of the most transformative blockchain use cases is a trusted supply chain management, which can provide an agreed-upon, shared record of the asset information across the entire supply chain network. Manufacturing provenance is an extension of this use case where blockchain holds complete provenance details of each component and makes it accessible by manufacturers and other stakeholders.

Contract management letters of credit and commercial paper management are also related use cases in which shared ledger technology is used to monitor a shared repository of legal documents with their approval histories. The business process or workflow of document handling is enabled by chaincode, which automatically checks the consistency of these documents to reduce errors.

Blockchain can also verify governmental functions like vehicle registration, livestock registry, food safety provenance, and equipment maintenance. Furthermore, healthcare verifications and payment records can also be matched, reviewed, verified, approved, and paid. Blockchain is also the platform upon which cryptocurrency is built.

E-commerce Function

Cloud-based e-commerce supports enhanced customer engagement as well as supplier-and-partner engagements. The core transformation components (marketing, customer analytics, and e-commerce) can enable a human-like enrichment designed to delight while giving the right experience at the perfect moment that builds lasting loyalty. One critical determination that must be made is deciding if current on-premises components should be deployed onto a cloud service provider platform. Resilience and elasticity are among the primary reasons for opting to use cloud-based services.

If a PaaS offering is selected, many architecture elements are available as part of the platform, and only configuration and deployment are required. If a SaaS solution is chosen, management is reduced to only configuration and user management.

The e-commerce architecture (figure 2.5) applies to digital transformation initiatives within most industries. The deployment considerations vary with organizational maturity, investment plans, customer expectations, specific products and services, and elasticity or availability requirements. The cost and effort required to implement PCI or data security compliance frameworks should also guide decisions on what cloud deployment and services model to adopt.

An e-commerce cloud deployment requires first selecting the required business model capabilities and goals. Many of these business models include the use of retailers to deliver a supplier's product or service. When doing this, retailers often provide both payment processing and payment gateway services. Merchandising and marketing to support the customer's buying journey are also helpful additional services. Extended multichannel communications for these activities could include email, games, and social media. Delivering a digital experience and customer care support across the transaction lifecycle are also excellent adjuncts to e-commerce. The delivery of personalized care requires access to a broad range of data hosted on many disparate systems.

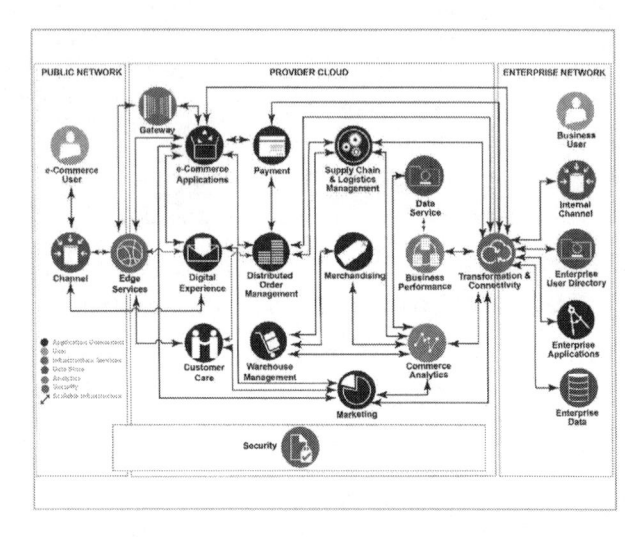

Figure 2.5. E-commerce platform functional components.

From a logistics-and-supply-chain point of view, common e-commerce requirements include supplier/retailer system communications, warehouse management, and distributed order management. Data services can also be used to generate and aggregate supply chain and logistics reports.

Enterprise Social Collaboration Capability

Enterprise social collaboration (figure 2.6) enables organizational engagement across employee, customer, and partner interactions. Social collaboration solutions in different industries align key business initiatives. Their value is in helping different roles or departments meet their business needs within the context of an overall collaboration initiative and organizational goal. Social collaboration advances digital transformation goals and benefits key business roles and functions through the enablement of business patterns and social business scenarios.

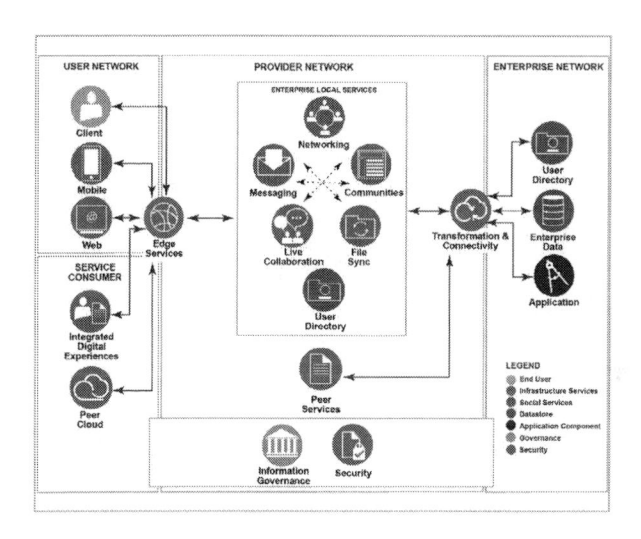

Figure 2.6. Enterprise social collaboration.

The key to delivering value is the provisioning of a collaborative information exchange with intelligent and secure social applications. With this, users experience an integrated ecosystem of collaboration and communication applications and services while maintaining access to underlying services that can be infused into business processes, integrated with other applications and services, and aggregated into other experiences. Multiple business needs can be met by applying social technologies to enable collaboration and process integration within the enterprise and across an organization's entire ecosystem.

Internet of Things Function

The Internet of things (IoT) links physical entities to information technology systems (figure 2.7). Data derived from this linkage is used to inform multiple applications and services. Use cases for this span across multiple applications, enterprises, governments, and consumers. Electronic devices are fundamental to IoT because they interact with the physical world, sensors that gather information about activities (objects and humans), and actuators that can act on objects. Fifth generation telecommunications systems (see Chapter 5) will cause an explosion of IoT in the near future.

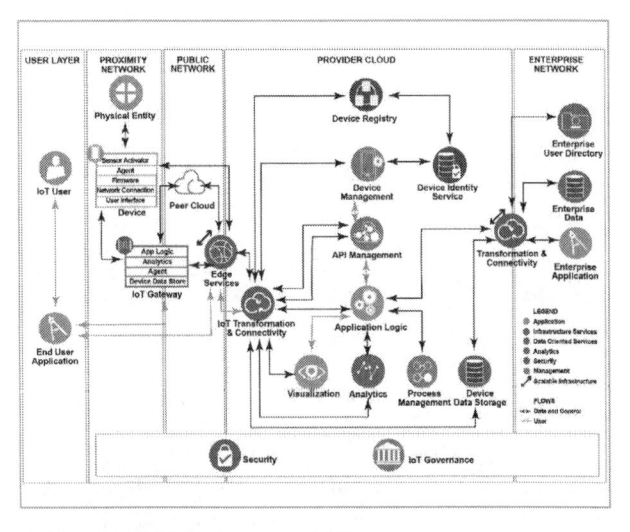

Figure 2.7. Internet of Things components..

The cloud delivers the edge, platform, and enterprise components of IoT. The edge collects data from endpoint devices, transmits that data to devices, and links to the back-end data center on a limited basis. The platform receives, processes, and analyzes edge data. It also delivers API management and visualization services. Transformation and connectivity module manages dataflow across the enterprise network, directing it to an enterprise data lake. Governance and security subsystems ensure the enforcement of security controls and policies. Regulatory compliance is also tracked.

Data involved in an IoT system may include personally identifiable information (PII). This would require additional attention to legal and regulatory compliance requirements. Devices may be directly linked with individuals, or specific persons could be the subject of sensor-provided data. Caution must also be taken if aggregated data can be used to identify the related person. A residential electricity meter is an example because its readings can be directly related to the individuals who live in the home. PII protections may involve where and how data can be stored, the information owner, and any required data usage restrictions. Data protection requirements can have a wide range of implications.

Two areas of rapid IoT growth are healthcare and transportation. An example of a healthcare implementation could be an IoT system support of a diabetic patient through the use of a continuous glucose-monitoring device, an insulin pump, and activity-monitoring devices on the patient's body. In the transportation industry, connected vehicles require real-time event detection and management systems designed to detect, analyze, and securely handle events.

Mobility Function

Modern business and mission models require an ability to provide lifecycle support to enterprise mobile applications and devices deployed to the employee, partner, and customer devices. The enterprise must also provide managed access to back-end business applications and enterprise data sources that support mobile applications and devices (figure 2.8).

Cloud computing and cloud services are perfect for supporting the time variable usage patterns of mobile apps. The scalability and elasticity of cloud computing can vary back-end resources to match increasing and decreasing levels of mobile-device requests. Mobile applications also make use of app unique, server-side data. Non-enterprise data, such as social media data, can also be incorporated.

Data associated with mobile apps often have access and volume requirements that are difficult to meet with traditional transaction-based enterprise systems. Mobile apps are also commonly supported using multiple databases. Such databases can hold necessary enterprise data in a form suited to mobile apps. The elastic provision and support of these application-specific databases is a cloud computing strength. The use of application-specific databases also reduces enterprise systems access requirements and associated resource needs.

Mobile capabilities can be deployed using any of the three core cloud computing service models. With IaaS, the CSP supplies underlying compute nodes and data storage. The customer is responsible for designing the architecture, installing the required software, and configuring them to work as a cohesive whole. When PaaS is consumed, the provider offers most of the mobile service components as a bundled set. The customer selects and configures the needed services and develops and required custom code.

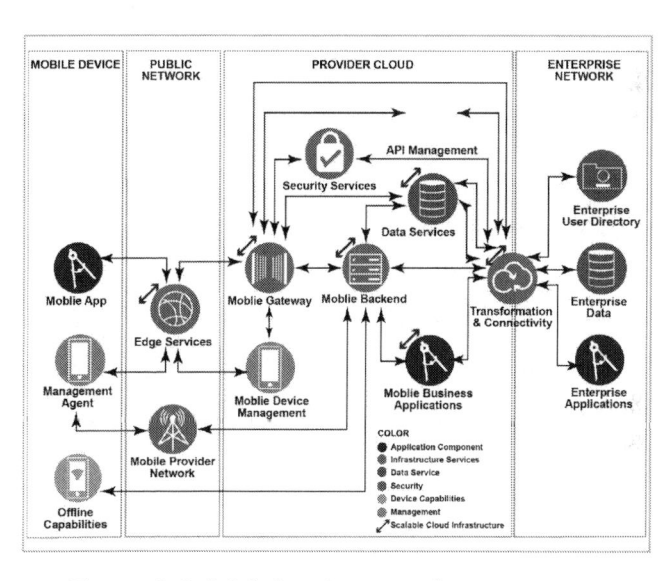

Figure 2.8. Mobility functional components.

Secure Data Management Function

Data security is the predominant issue whenever enterprise cloud computing use is discussed. Developing and deploying business solutions using cloud services, however, requires a clear understanding of the available security services, components, and options (figure 2.9).

Key aspects are establishing a consistent way to manage identities and access to platforms and applications from everchanging and unknown entities (human and non-human). Tenant isolation across the compute, data, and networking infrastructure must be ensured. This includes safeguards against application and network threats, exploits, and vulnerabilities. Organizations must provide secure connectivity to data at the enterprise and protect sensitive data in transit, at rest, and in use in the cloud. Visibility into virtual infrastructures should be maintained by collecting and analyzing data in real time across the various cloud components and cloud services. Optimizing the methods, processes, and tools for running security operations is key to keeping the overall cost low. The enterprise must consistently assess security practices, plans, and designs and promptly evolve them to stay ahead of threats.

If SaaS is a component of your business model, the cloud service provider typically takes most of the responsibility for security. This doesn't alleviate the enterprise entirely, though, since the organization retains responsibility for identity management and service authorization governance. PaaS services dictate CSP as the chief protector for security aspects of the software which makes up the platform, including the operating system and any middlewares and runtimes made available for use by the customer. Using the IaaS service model, the cloud service customer takes a significant amount of responsibility for the security of data, applications, systems, and networks. Roles and responsibilities relating to cloud services information security must be clearly defined. Information security responsibilities are shared between the cloud service customer and the cloud service provider. This responsibility split varies based on the technology service being consumed.

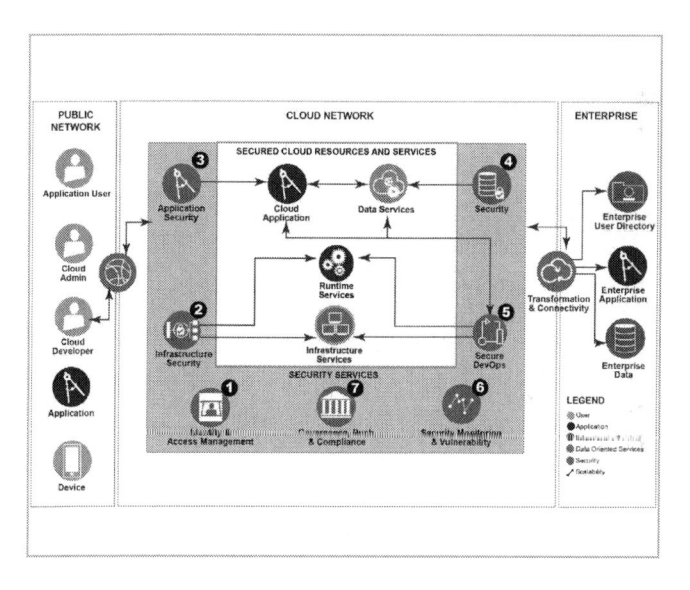

Figure 2.9. Secure data management functional components.

Organizational security policy plays a pivotal role in determining how the organization's IT systems achieve security goals. When adopting cloud services, the security policy must extend to include cloud service security policies that take into account the different environments involved in using cloud services. The policy must also address the three major cloud service models—IaaS, PaaS, and SaaS.

An organization's digital transformation goals, business strategy, competitive differentiation, and industry regulations are prominent factors in shaping corporate data management and security strategy and governance. Security governance must be enforced through the use of security controls. The education of employees and end users on security guidelines is imperative. Policies and procedures must be designed to meets industry and compliance regulations while maintaining operational efficiency. A continual assessment of risks, especially those associated with suppliers and sub-processors, is key.

The security controls are generally deployed across various infrastructure layers. These layers include identity and access management, data, applications, network and server infrastructure, physical security, and security intelligence. When adopting cloud services, enterprise security policy must be updated to define how enterprise controls interact with those provided by the CSP. Cloud service customers should seek information from the CSP on their cloud security policy and the security controls. Only with this information can the customers ensure that their use of the cloud services can meet the requirements of the customer's cloud security policy.

Summary

In Chapter 1, I discussed how cloud computing has made technology irrelevant. Henry Ford did not need to know the ins and outs of an engine. The standards he set for the engine allowed Ford Motor Company to operate on a much larger scale.

Replace engines and radiators with the functions and capabilities, and you can see the future of IT. Hybrid IT strategies can transform the world in ways that Ford transformed the automotive industry. Technical knowledge no longer qualifies you to start a company that disrupts an entire industry. I learned this firsthand, as the founder of SourceConnecte.

SourceConnecte is a digital supply chain company. It began when a colleague asked me if I had knowledge of supply chain. My answer? "No." The conversation could have easily ended there, but I didn't need to know supply chain to help build a supply chain company. All I needed to know was the business processes involved in supply chain. I knew blockchain. I knew cloud computing. Most importantly, I knew how to gather software involved in supply chain and build a hybrid IT strategy.

My first position within SourceConnecte was that of a consultant. After identifying key business processes, I found the right technology to help the process move along. Sometimes, this only took a Google search. Other times, I reached out to colleagues who could connect me to the right technology. It didn't matter that I, my colleagues, or anyone else in SourceConnecte didn't know the bits and bytes involved in every software.

WordPress powers the SourceConnecte website. Inxeption takes care of e-commerce. IBM Sterling handles inventory. My initial job with SourceConnecte was not only to discover which APIs we should use, but also how to connect them to accomplish larger business goals. Integrating these APIs allows SourceConnecte to give visibility to companies who are in the transactional stage of growth. We would have never been able to reach these goals if it weren't for the access and standardization of software that allowed us to digitize the supply chain.

A simple conversation with a colleague quickly grew into a digital supply chain company. We did everything in three months. How much money did we spend in those three months? Zero dollars. Did we create any of the IT infrastructures? Not at all.

How easy was it to transition from the role of "consultant" to that of "founder?" Very easy. Even though I had no prior knowledge of supply chain, I understood how to seek out, connect, and use APIs to accomplish the goals of a supply chain company. As you'll learn in future chapters, this was not done unintentionally. Past failures taught me the importance of prioritizing processes over technology. By focusing on specific processes, I was able to realize our end results without hiring a single employee or spending a single dollar.

3

Transformation Business Models

Cloud-based business enablers that power today's transformative business models are driven by the capabilities and functions described in Chapter 2. The business models are transformative because of their ability to take advantage of platform economics. After explaining the genesis of key hybrid cloud business enablers, Chapter 3 demonstrates how they can be utilized to establish an organization's customer value proposition and a competitive strategy that effectively exploits hybrid cloud computing benefits for digital transformation.

Hybrid Cloud Business Enablers

In 2012, the IBM Institute for Business Value did a study on how cloud computing was affecting business innovation and strategy.[5] The institute surveyed 572 business and technology executives worldwide to determine how organizations used cloud computing then and how they planned to use it in the future. Although the research indicated full recognition of the cloud as an essential tool, relatively few organizations actively embraced the cloud as an option for driving business model innovation. The survey did, however, indicate that a dramatic change in that perception would occur over the next few years as more organizations were looking to cloud to drive new business and transform industries. The research also identified game-changing business enablers that were powered by the cloud.

Six years later, in 2018, the institute again looked at the business value of cloud computing. This second study[6] surveyed 1,106 executives across nineteen industries and twenty countries. The research not only confirms the earlier findings, but also revealed that 85 percent of companies surveyed were already multi-cloud environments, most of which were multiple hybrid clouds. Around 76 percent of organizations reported that they were already using two to

---

[5] The power of cloud: Driving business model innovation http://www.ibm.com/common/ssi/cgi-bin/ssialias?subtype=XB&infotype=PM&appname=GBSE_B_TI_USEN&htmlfid=GBE03470USEN&attachment=GBE03470USEN.PDF

[6] Assembling your cloud orchestra: A field guide to multi-cloud management, https://www.ibm.com/downloads/cas/EXLAL23W

fifteen clouds, and 98 percent forecast they would use multiple hybrid clouds within three years.

Here's more data:

- 49 percent were establishing a multi-cloud architecture to develop new and enhanced products and services.
- 46 percent indicated a need for a multi-cloud environment to support agile business processes.
- 51 percent were using multiple clouds to cultivate a flexible, modular infrastructure that could quickly absorb and leverages technological advances.
- 62 percent required multiple clouds to create innovative business models.
- 52 percent needed multiple clouds to produce new revenue streams.
- 66 percent need them to enhance margins.

These wide-ranging and valuable potential advantages strongly suggest multi-cloud environments as essential to survival and success in today's digital era.

A 2018 study results more than verified the earlier studies list of six key observable cloud computing business enablers (figure 3.1):

- Cost flexibility
    - Shift fixed cost to variable cost
    - Allows "pay as and when needed" model
- Business scalability

- - Provides flexible and cost-effective computing capacity to support growth
- Market adaptability
  - Enables faster time to market
  - Supports experimentation
- Masked complexity
  - Enables expanded product sophistication
  - Allows for more end-user simplicity
- Context-driven variability
  - Enables user-defined experiences
  - Increases product relevance
- Ecosystem connectivity
  - Fosters new value nets
  - Drives potential new businesses

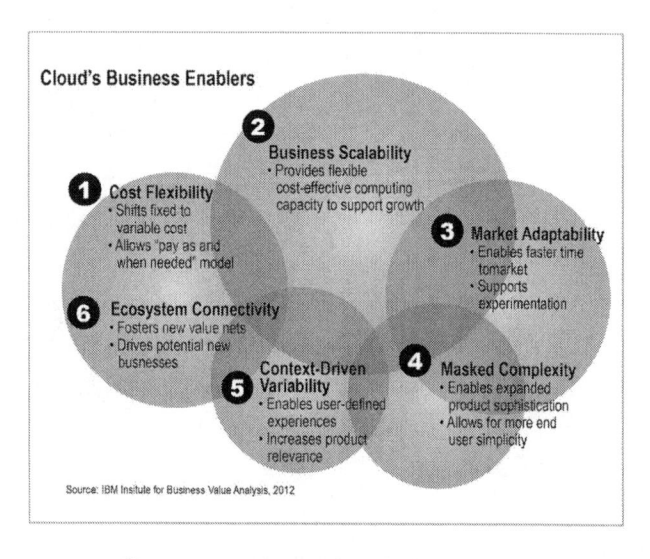

Figure 3.1. Key cloud computing business enablers.

An example of some of the additional benefits that can logically extend from these include the following:

- Blending cost flexibility with business scalability that addresses business growth as well as business contraction
  - Provides significant cost reduction with necessary IT services available at a fraction of the cost of traditional IT services with upfront capital expenditures that eliminated and dramatically reduced IT administrative burden
- Blending business scalability with ecosystem connectivity
  - Provides increased flexibility by using on-demand computing across technologies, business solutions, and large ecosystems of providers, all the while enjoying reduced new solution implementation times
- Blending masked complexity with market adaptability
  - Provides access to the organization's digital products and services anywhere at any time or any device to any customer segment
- Blending context-driven variability with ecosystem connectivity
  - Significantly enhanced service quality that drastically improves customer experience through highly reliable and timely service delivery

- Blending ecosystem connectivity with masked complexity
  - Provides rapid delivery of personalized composite digital products and services while simultaneously increasing the profit margin on core products and services
- Blending market adaptability, cost flexibility, and ecosystem connectivity
  - Provides accelerated innovation and ability to capitalize on new revenue opportunities through rapid collaboration, crowdsourcing, and cocreation

The second study also revealed that even though an overwhelming majority of enterprises were operating multi-cloud architectures, few had effective management control of those environments. Even with 98 percent of organizations expecting to embrace multi-cloud architectures by 2021,

- 41 percent have a multi-cloud management strategy;
- 38 percent have the procedures and tools in place to operate a multi-cloud environment;
- 30 percent of enterprises have a multi-cloud orchestrator or multi-cloud management platform that can choreograph workloads; and
- Less than 40 percent of organizations have cloud configuration management tools that provide information about resource configuration and relationships between resources.

For the attainment of business goals, the most important findings were that enterprises using multiple clouds outperformed their competitors in multiple vital metrics, including revenue growth, profitability, efficiency, and effectiveness in achieving business objectives.

The valuable business enablers identified and verified by these two IBM cloud computing studies are themselves enabled by transformation capabilities and functions presented in Chapter 2, for example:

- The ability to embed cost flexibility in any business model requires an ability to recognize critical process indicators (big data analytics) and a means for managing business process interfaces (API management).
- Identifying when to scale a business process up or down (business scalability) while simultaneously protecting any related intellectual capital requires secure data management and management of the APIs (API management) that connect different business processes.
- Market adaptability requires an ability to reach customers on their mobile devices (enterprise mobility), effective communications between business employees and customers (enterprise social collaboration), and the ability to execute financial transactions digitally (e-commerce).
- The ability to mask the complexity of delivering digital services to remote customers requires the ability to securely manage mul-

tiple types of end points devices (Internet of things) over a wireless network.
- Leveraging an ability to tailor a digital service to a customer's current environment (context-driven variability) requires the collection and analysis of data from the remote environment (big data and analytics) and to communicate with that customer.
- Interacting with a business partner to jointly deliver product and services (ecosystem connectivity) requires API management for managing business process interfaces, blockchain to ensure the immutability of data records, and secure data management.

The final challenge to developing a transformative product or service is an organization's ability to link a properly managed multi-cloud environment that is capable of delivering applicable transformation capabilities and functions to service the desired cloud business enablers using a business model that has proven itself to be compatible with operations based on platform economics, namely:

- Freemium
- Ad revenue-based
- Customer service differentiation
- Public service
- Product pairing

Companies that have been successful in operationalizing this explicit linkage from a multi-cloud environment to the effective running of a platform economics friendly business model are revolutionizing industry verticals. Transformative business models exploit this linkage in taking advantage of platform economics by delivering products and services in a manner that delivers value to a distinct customer segment.

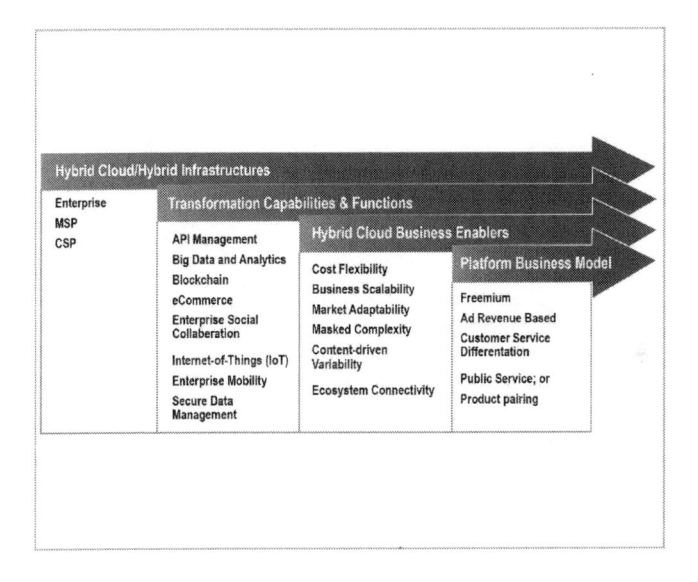

Figure 3.2. Creating a transformative business model.

Customer Segmentation

The next link in this success chain is the identification of a profitable customer segment. Success is dependent on a thorough, accurate, and continually re-evaluated understanding of the chosen segment. This is a dynamic marketplace-sensing process that cannot be replaced by any static dataset. Often described by terms like *personalization*, *customization*, and *individualized*, this requirement highlights the truism that successful business models are founded upon intimate customer knowledge. Widely used static marketplace segmentation models include the following:

- Geographic
  - It segments the market based on a pre-defined geographic border.
- Demographic
  - It divides a market through publicly available characteristics variables such as age, gender, education level, family size, occupation, and income.
- Psychographic
  - Segments based on intrinsic traits such as personal values, personalities, interests, attitudes, conscious and subconscious motivators, lifestyles, and opinions. Understanding target customers on this level normally uses methods like focus groups, surveys, interviews, and case studies.
- Behavioral
  - Uses specific reactions and the customer's decision making and buying processes as a basis for market segmentation. Attitudes towards a brand or the way a product is used are examples of behavioral segmentation.[7]

---

[7] https://learn.g2.com/market-segmentation

Most brands use more than one segmentation technique, and many use a combination of several at once. In recent years, these traditional customer segmentation approaches have been gradually abandoned in favor of new, data-driven approaches. This new model has delivered improved efficiency and consumer relevancy.[8] Customer journey mapping is leading the way in marketing-based analytics as it allows customer experience to be personalized for each individual. These behaviors act as pointers to the most appropriate strategy for presenting your relevance to that segment, convincing segment decision-makers to buy your product or service and capturing market share within the selected industry.

Segment Strategy

By continually collecting and reviewing data on customer segment purchasing behavior, enterprises can select the most appropriate path toward making data profitable. Modern data-centric business strategies as succinctly outlined by Q. Ethan McCallum and Ken Gleason in *Business Models for the Data Economy* are the following:

- Collect/supply
- Store/host
- Filter/refine
- Enhance/enrich
- Simplify access

---

[8] https://ieeexplore.ieee.org/document/8392838

- Obscure
- Consult/advise

Collect/Supply

This business model gathers and sells raw data. Data is manually gathered or automated by scraping websites and then sold to interested parties. This option leverage scalability by selling the same dataset to multiple clients and customer segments. Although startup costs may be high, there is a near-zero cost with electronic distribution. The highest recurring cost is in storage and bandwidth, which should reduce over time. The efficiency and simplicity of this option are at the heart of its popularity.

Store/Host

With store/host, scalability is used to reduce the cost of bulk storage. That cost reduction is leveraged to provide margins in offering storage service to others. Although some large organizations can leverage the volume storage route themselves, smaller companies or those that don't see IT as a core capability often choose to off-load management to third parties. This is also very useful for extensive datasets or otherwise difficult to manage datasets.

As a service provider, you take on that burden. A store/host strategy can be especially profitable when dealing with broadening international data privacy and protection regulations. The legal and contractual burdens may be significant, but clients pay well to be relieved of the burden.

Economies of scale work in the provider's favor when the marginal cost to store data falls below the marginal value of each client consuming data storage. Hosting isn't limited to storing and providing access to raw data. Basic summary statistics, calculated measures of the datasets, and other analytical services can also be provided. This option provides value by eliminating the clients' need to download the data for self-analysis.

Filter/Refine

More interesting data subsets can be created by stripping out problematic records or data fields. Similar to the collect/supply strategy, this option adds value by handling the technical data cleansing work. This can be an especially valuable approach when contact data is part of a solution. In this case, clients pay for data duplicate removal, verification, or consolidation services. Normalization and downsampling are two other important refinements that could be offered.

Enhance/Enrich

Blending two or more datasets to create new or original perspectives is the goal of this business model. The goal is to spare clients the burden of preprocessing data themselves. The strategy focuses on adding information instead of normalizing or removing data. A unique value proposition can be created by joining datasets or offering computationally intensive processing services on single datasets.

This enhanced data can then be offered to the market with a profit. Combining datasets in a logically intuitive but may be challenging to perform due to the data structure or location could also be profitable. Public domain and other open datasets are ideal candidates for enhance/enrich operations.

Simplify Access

Simplifying data access helps clients easily pull data in a preferred format. Many companies do not want to manage bulk data downloads, especially when the data is only available in formats that are not compatible with modern machine-readable interfaces. This strategy is a logical extension to a collect/supply or filter/refine business. Scalability could be easily leveraged by offering the same raw dataset with identical or similar postprocessing.

Data could also be hosted behind an API and made available for programmatic access to specific subsets in a machine-readable format. Clients who wish to consume large datasets either download all the data and write their extraction routines or pay for a service to subset and extract on demand. The ability to extract specific subsets of data can be just as valuable as having the dataset in its entirety. Most of the effort in data analysis is spent on the collection, segmenting, subsetting, and cleaning processes.

Obscure

Inhibiting people from seeing or collecting certain information can be very profitable. There are also many business opportunities in keeping data inaccessible. Companies wish to protect their data-in-transit against exploitation of their data exhaust or similar data byproducts. Expanding national data sovereignty and individual data privacy concerns make protection against data scraping and aggregation efforts increasingly profitable. Enterprises can leverage the dichotomy by building tools to obscure information or otherwise foil data collection. Businesses go to great lengths to keep data inside the corporate walls through the use of VPN access and internal websites.

Consult/Advise

Consulting provides guidance on the data efforts of others. This business model is one of the most open-ended strategies. While not unique to the data arena, it does offer the opportunity for profit generation in the advising firm if they can uniquely process existing datasets. A consulting effort could also crosscut other data analytics and obfuscation strategies. Consultants trade experience and expertise for money. One critical element of a data consulting operation is domain knowledge beyond that of the client's industry.

These strategies are at the heart of transformational business models because selecting the right business model maximizes customer interactions, revenue-based transactions, and pave the way toward enhancing margin, profit, or budget utilization. As presented earlier, business models that have proven themselves to be compatible with operations based on platform economics are these:

- Freemium
- Ad revenue-based
- Customer service differentiation
- Public service
- Product pairing

Transformative business models target the right customer segment with the most appropriate strategy and deliver new and innovative customer value propositions.

Segment, Strategy, and Business Model Alignment

Analysis of the 2018 IBM IBV study revealed that organizations were harnessing the cloud to recast customer relationships and transform both product and service development. The best three examples of how organizations were using the cloud to impact company profits, industry value chains, and customer value propositions were as follows:

- Optimizers use the cloud to enhance customer value propositions incrementally and improve organizational efficiency.
- Innovators improve customer value through cloud adoption that results in new revenue

streams or even changing their role within an existing industry ecosystem.
- Disruptors rely on the cloud to create radically different value propositions, as well as generate new customer needs, segments, and industry value chains.
- Businesses were relying on cloud computing to enhance internal efficiencies and target more strategic business capabilities. In 2012, the number 1 objective for adopting cloud was increased collaboration with external partners. Only one of the top seven objectives focused on internal efficiencies.

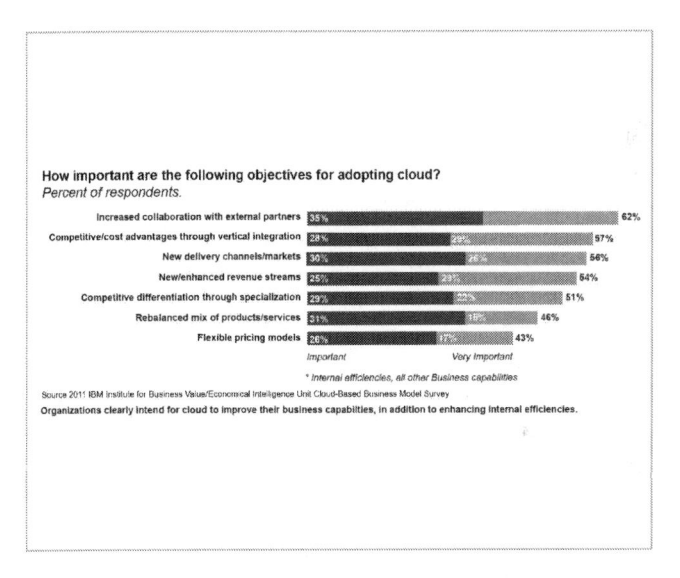

Figure 3.3. Top cloud computing business objectives.

The successfully designed transformative business models will deliver a product or service capable of fulfilling one of the following customer value propositions to the selected customer segment:

- *Enhance.* Organizations can use the cloud to improve current products and services and enhance customers' experiences. This option also retains current and attract new customers, garnering incremental revenue.
- *Extend.* The cloud can help create new products and services or use new channels or payment methods. This attracts existing or adjacent customer segments. This approach can generate significant new revenues.
- *Invent.* Companies can use the cloud to create a new "need" and establish leadership in a unique market. This attracts new customer segments and generates brand-new revenue streams.

These customer value propositions are being used within industry vertical to fundamentally alter industry value chains in one of the following manners:

- Improve how an organization maintains value chain position through increased efficiency and an improved ability to partner, source, and collaborate (improve)
- Transform a company's role within its industry or enter a different industry by assisting in developing new operating capabilities (transform)

- Use the cloud to build a new industry value chain or disintermediate an existing one, radically changing industry economics (create)

Value chain modification shifts which entity creates value and how it is created, delivered, and captured. When correctly linked to a digital transformation strategy, hybrid cloud computing and hybrid IT can ensure tight alignment that targets a profitable business model, across a targeted customer segment that also aligns with an appropriate competitive strategy. Identifying the intersection of customer value proposition and value chain modification targets maximizes the probability of attaining desired organizational goals. This approach enables the active exploitation of key hybrid cloud computing benefits for a successful digital transformation.

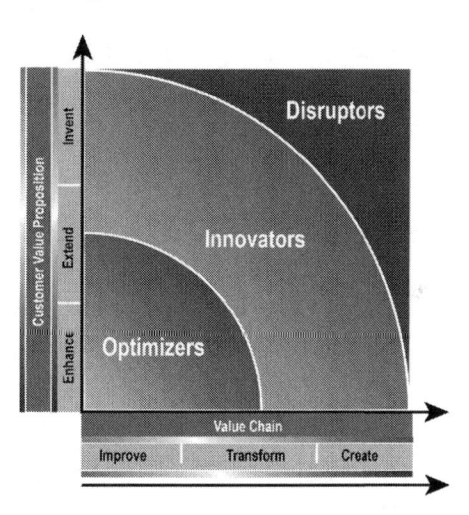

Figure 3.4. Cloud enablement framework.

Summary

We benefit from disruptors and innovators every day. Uber, Facebook, and YouTube all use cloud technologies to enhance customer services, extend customer segments, and invent new revenue streams.

Let's take a look at a disruptor that I have used frequently for both personal and business travel: Airbnb. Airbnb uses not one but two customer segments as data sinks. One segment is looking for private living spaces. The other is looking for additional income. Airbnb can also use each customer segment as an information source. Customers need private living spaces, and property owners provide the information necessary for customers to make their selection. Property owners need people to book their spaces, and customers offer their data, useful to both Airbnb and the property owners using the site. Airbnb also created new revenue streams by allowing property owners, hosts, and travelers to connect in new ways.

It's that simple. Identify data sinks and data sources. Use cloud computing to connect them.

Cloud computing does not give people the idea to make these connections; it just makes these connections possible. Airbnb would not exist without optimizers, innovators, and disruptors leading the way. The human brain will forever be the most powerful tool involved in any transformational innovation.

Optimizers, innovators, and disruptors all share a passion that I have recognized throughout my career. Even before entering the workforce, I looked for ways to find solutions through connection. I have been able to apply this passion for over forty years, in ways that I could have never thought possible.

As a child, I recognized that I didn't need a computer to bring ideas, services, and products together. Neither did Henry Ford, Bill Gates, or Howard Schulz. We have all harnessed the same process of identifying relationships and making connections between disparate domains. This was merely a passion for many years—not a way to build a business. It wasn't until the excitement of Y2K and the dawning of the Information Age that I started to truly understand how this passion could be transformed into power. Creating relationships and connections rather than the IT services themselves were the key to innovation.

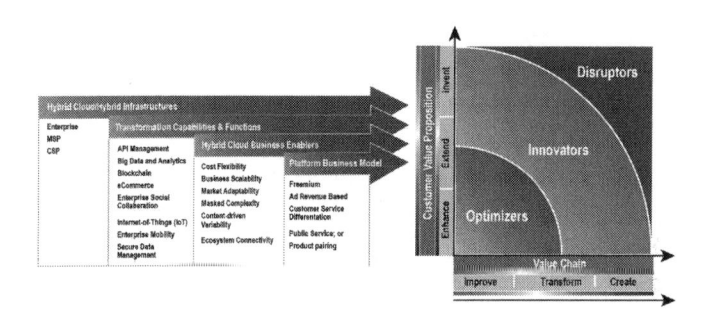

Figure 3.5. Transformative business model support organizational customer value proposition.

The world has transformed immensely since the time I began working in technology, but one thing remains the same: an open mind, an eye for data sources and sinks, and the ability to connect can turn any business blueprint into a working reality.

4

Transformation Infrastructure

Hybrid IT enables a composable infrastructure which describes a framework whose physical compute, storage, and network fabric resources are treated as services. Resources are logically pooled so that administrators need to physically configure hardware to support a specific software application, which describes the function of a composable architecture. Chapter 4 describes and explains why composable architectures have become essential to the modern enterprise. It also examines how they enable the transformative business models outlined in Chapter 3.

The value of composable infrastructures lies in their ability to enable enterprise agility. It can also preserve and optimize both economic and operational return on investment in IT. The technical complexities of changing to a technology consumption model can also be eased. From a management viewpoint, the use of composable architectures transforms the CIO and IT function into the collaboration and integration hub across all enterprise functions. Cloud has put the fast forward on innovation and now technology infrastructure is becoming a business commodity; the modern CIO/IT department should evolve as a broker of services instead of a purveyor of technology infrastructure to the business units.

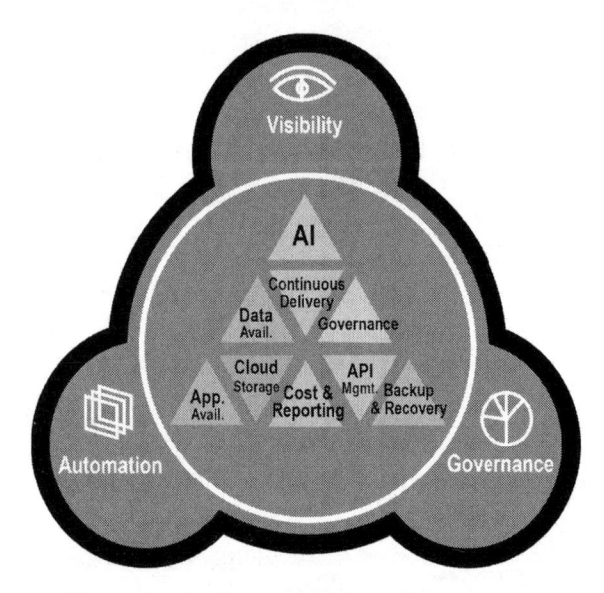

Figure 4.1. Composable infrastructure.

This type of transformative infrastructure is foundational to contemporary agile business because a hybrid IT environment, private clouds, public clouds, community clouds, traditional data centers, and services from service providers must be integrated and interconnected. The agility afforded by architecture composability lets applications and services be deployed to, and consumed from, the most appropriate service environment combination. When public clouds are used as an architecture component, enterprises are able to scale continuous delivery and innovation into product and service specifications.

Composable architectures have proven themselves to be an effective solution for many firms undergoing a digital transformation journey. To follow this path, enterprises also implement an IT infrastructure design and optimization strategy that focuses on application portfolio rationalization.

Composable Infrastructure

From the infrastructure side, this modernization task transforms legacy data centers into private clouds and migrates existing legacy or packaged applications onto this highly automated environment. This initial step toward establishing a hybrid cloud environment also enables a rational and collaborative adoption of public cloud infrastructure services (IaaS). It reduces the friction often caused by retraining staff in public cloud operations, modern infrastructure technologies, and composable solution management tools. Composable infrastructures can build new revenue-generating products and services faster while simultaneously addressing the key inhibitors to change, which include the following:

- General concerns regarding lack of adequate hybrid infrastructure security
- The false impression that cloud cannot support the operational/performance requirements of critical applications (e.g., SAP and Oracle)

- Management challenge presented by multi-cloud environments contracts that will include varying levels of governance and service-level agreements (SLAs)
- The need to match employee management skills across various cloud platforms

Composable infrastructure architectures have two major functions. They must be able to disaggregate and aggregate resources into pools and compose consumable resources through a unified API. Overseeing these functions is management software that can also communicate with the API management functions reviewed in Chapter 2.

Continuous Delivery

Composable infrastructures also enable the implementation of all IT software, infrastructure, and security features into production safely and quickly in a sustainable way. Broadly referred to as "continuous delivery," this capability can significantly reduce the risk often associated with software deployments by establishing an operational environment within which updates and deployments can be performed on demand. The blue-green deployment technique is often used for this as it reduces downtime risks by running two identical production environments. Only one environment is "live" and serving app production traffic at any instant.

Since all IT teams work together in a collaborative environment to deploy onto a fully automated infrastructure, the enterprise enjoys a faster time to market. Automation helps avoid large amounts of application development rework by infusing regression testing into the daily workflow. Higher software quality lets developers dedicate more attention to research, exploratory testing, usability, performance, and security.

The resultant CI/CD deployment pipeline ensures that quality is built into all enterprise products and services from the outset. This subsequently reduces the cost of incremental changes and improves the economic viability of small-batch development, A/B testing techniques, and hypothesis-driven approaches to feature development.

Multi-Access Edge Computing

Fifth-generation (5G) wireless networks will significantly enhance the current mobile network environment. These new networks will use multi-access edge computing (MEC) to extend composable enterprise infrastructures to the network edge, a capability broadly referred to as edge computing.

By using edge computing, content and applications can operate apart from their core data center in a disconnected manner. In edge computing, the data is processed on the device or sensor itself without being transferred to a dedicated compute resource. A similar concept, fog computing, describes an optional approach where the data is processed within a fog node or IoT gateway that is situated within the local area network. This approach delivers services characterized by ultra-low latency and high bandwidth and real-time access to mobile network information. The network information can subsequently be used to provide prioritized application services. Composable enterprise infrastructures need to integrate and operate with geographically distributed compute and storage resources. To support this future IT-operating environment, enterprise content and application developers need to collaborate with telecommunications network operators to gain access to edge services. This allows more enterprise flexibility and speed in deploying innovative edge computing applications and services. Chapter 5, "Transformative Networks," will address these changes in detail.

Application Design

Most modern applications are designed based on the distributed computing software development model. This model uses a client-side to initiate server requests and a remote server-side to process these requests. With this client-server model, application developers leverage centralized compute and storage, which has been a major driver of cloud computing. For MEC, applications developers need to identify application features that require processing at the edge as distinct from features that need high compute power or that do not require near real-time response. Applications need to be designed to support distributed processing, synchronization of contexts, and multilevel load balancing. They should also leverage infrastructure as code (IaC), which describes the complete automation of infrastructure provisioning and deprovisioning.

The information technology ecosystem is quickly moving toward consuming cloud service provider services like Microsoft's Azure IoT stack, Greengrass for Amazon's AWS Lambda, and GE's Predix. Greengrass, for example, consists of the AWS Greengrass core for edge computing capabilities working with AWS IoT software development kit–enabled IoT devices. Using this architecture, AWS IoT applications can respond in real time to local events and use cloud capabilities for all other data processing functions. To deliver these new services and maximize the value of MEC, it is also important for the application developers and content providers to fully understand the core characteristics of the MEC environment and the distinguishing MEC services.

MEC also introduces a different standard for the edge computing paradigm. A MEC point-of-presence (PoP) is different than a traditional cloud PoP and may offer significant advantages to edge applications and services. This makes knowledge of the targeted edge environment a crucial application design aspect.

Edge computing application design development model, therefore, has three locations:

- Client
- Near server
- Far server (see figure 4.2)

An end-to-end IT service designed to operate in an IoT environment follows this model also but with different reference names or components:

- Terminal device component
- Edge component(s)
- Remote component(s)

The IoT architecture emphasizes the distribution of components.

The client can be a smartphone car, smart home, or industrial location that runs a dedicated client application. The near server is located at the edge level while the far server is located in the data center and considered remote. The model is new, and guidance on how to properly architect the model is still being developed. Modern development approaches (e.g., microservices), however, make adoption easier. Industry standards will help accelerate the application development for edge computing and MEC adoption. A MEC host (figure 4.2) contains a MEC platform and VMs or containers that hold the compute, storage, and network resources for edge applications. MEC offers a secure environment where applications may discover, advertise, consume, and offer services.

A key principle of the modern system design is that applications and networks should be completely agnostic to each other. In this design assumption, network conditions and topology characteristics were considered as an environmental input out of the programmer's control. With this assumption, the application passively adapted to the networking environment. This assumption is not valid in edge computing and many IoT industry use cases where network aspects are integral to the application design. A specific example is modifying application behavior to deliver a better user experience if network throughput limitations are encountered (e.g., adjusting the video stream compression ratio in response to throughput throttling).

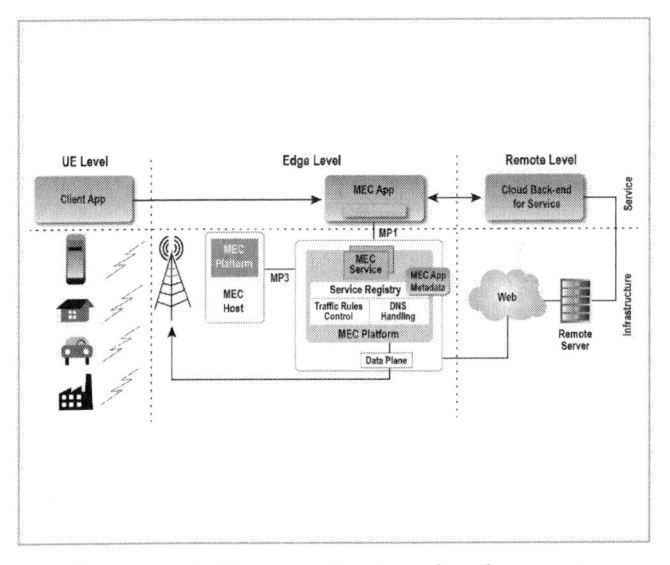

Figure 4.2. New application development paradigm introduced by MEC.

With MEC, the network and the applications operationally converge. While MEC can support any application and any application can run in MEC, MEC-aware applications can offer additional services. MEC application enablement[9] introduces such a service environment and can be used to improve the user experience. Using MEC, the network environment becomes less unpredictable, and contextual information can be used to dynamically adjust application behavior at runtime. This makes network characteristics a critical input to application design. MEC-aware applications, for example, can make a bandwidth request using the bandwidth management API to reserve networking resources in the MEC system. This could allow edge applications to benefit from low latency and high throughput in a predictable/controllable way. This additional environmental information can be leveraged at the time of service design to optimize the end-to-end service architecture.

A microservices-based architectural approach is well suited for MEC. This architecture leverages a different paradigm since an additional processing stage (at the edge) must be added to the application's workflow. An example would be doing preliminary processing in the edge device to determine the need for further action from core back-end services. This preliminary processing requires near-zero latency and needs the terminal device to support some computing capabilities (figure 4.3).

---

[9] https://www.etsi.org/deliver/etsi_gs/MEC/001_099/011/02.01.01_60/gs_MEC011v020101p.pdf

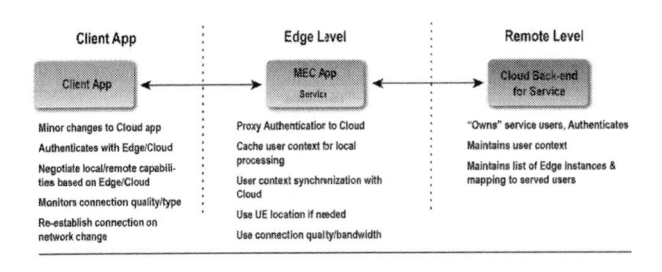

Client App	Edge Level	Remote Level
Client App	MEC App Service	Cloud Back-end for Service
Minor changes to Cloud app	Proxy Authentication to Cloud	"Owns" service users, Authenticates
Authenticates with Edge/Cloud	Cache user context for local processing	Maintains user context
Negotiate local/remote capabilities based on Edge/Cloud	User context synchronization with Cloud	Maintains list of Edge instances & mapping to served users
Monitors connection quality/type	Use UE location if needed	
Re-establish connection on network change	Use connection quality/bandwidth	

Figure 4.3. Example of splitting an application into "terminal," edge," and "remote" components.

MEC can implement computation off-loading techniques that can dynamically transfer processing between the terminal, the edge, and the remote component(s). This would support real-time adaptation to network conditions or improve application-specific KPIs, policies, or costs. The processing distribution may also be driven by user experience performance objectives. In the application design phase, the developer should consider how:

- a device application is to interact with the MEC system (service discovery, advertising, and consumption); and
- how an application may update the MEC data traffic routing rules.

With a choice to consume MEC services or produce them, the developer may want to consider additional service-related APIs.

MEP APIs

There are three important APIs to consider when developing a MEC aware application:

- Radio network information (RNI) API that provides radio network-related information that originates from the radio network based on industry-defined specifications. Typical information that can be exposed include the following:
  - Radio network conditions
  - Layer-2 measurement and statistics information
  - Information on users connected to the associated radio nodes
  - Changes to any previously provided information
- Location AP which performs active device location tracking or location-based service recommendations to allow a variety of additional services that are tightly coupled with a specific place (such as a shopping mall).
- Bandwidth manager API that allows a fair distribution of bandwidth resources between applications.
- User entity (UE) identity API that facilitates the association of IP traffic flows with a specific UE using an externally defined tag instead of the UE identity directly. This approach helps preserve user privacy by protecting identity information at both mobile and enterprise networks.

Application Trust Model

While 5G networks will create many new opportunities, they also increases the risk to enterprise data. As this new network will be heavily dependent on software-defined–network (SDN) technologies, data may be more vulnerable to the compromise of confidentiality by the interception of data communications or unauthorized access to a universally available set of network services. Since 5G will be the first cellular generation to launch in the global cybercrime era, the heavy funding by organized crime and nation states aimed at weaponizing cybercrime is a significant threat.

A software-defined network (SDN) is an information technology network that physically separates the network control plane from the data (or forwarding) plane in the data center. This technology is extended to wide area networks using SD-WAN (software-defined wide-area network) solutions. With this network management approach, a control plane controls several data plane devices. SDN uses network control plane software to enable dynamic, programmatically efficient network configuration in order to improve network performance and monitoring.

Replacing traditional dedicated hardware with a general-purpose computer and software is referred to as network function virtualization (NFV). In this environment, network services (i.e., routers, firewalls, load balancers, XML processing, and WAN optimization devices) are replaced with software running on virtual machines. These service functions become virtual network functions (VNF).

SDN seeks to separate network control functions from network forwarding functions while NFV seeks to abstract network forwarding and other networking functions from the hardware on which it runs. SDN abstracts physical networking resources—switches, routers, and so on—and moves decision making to a virtual network control plane. In this approach, the control plane decides where to send traffic while the hardware continues to direct and handle the traffic.

An SDN can control a traditional network with dedicated hardware or create and control a virtual network with software.

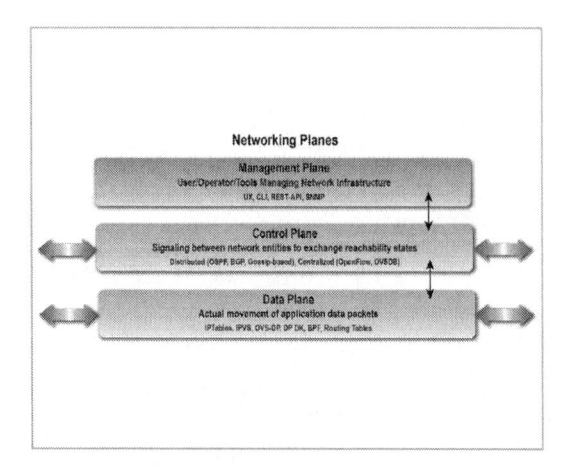

Figure 4.4. Networking planes.

Key cybersecurity tasks include the following:

- Securing the controller as the centralized decision point for access to the SDN
- Protecting the controller against malware or attack

- Establish trust by protecting the communications throughout the network by ensuring the SDN controller, related applications, and managed devices are all trusted entities
- Creation of a robust policy framework that establishes a system of checks and balances across all SDN controllers
- Conducting forensics and remediation when an incident happens in order to determine the cause and prevent reoccurrence

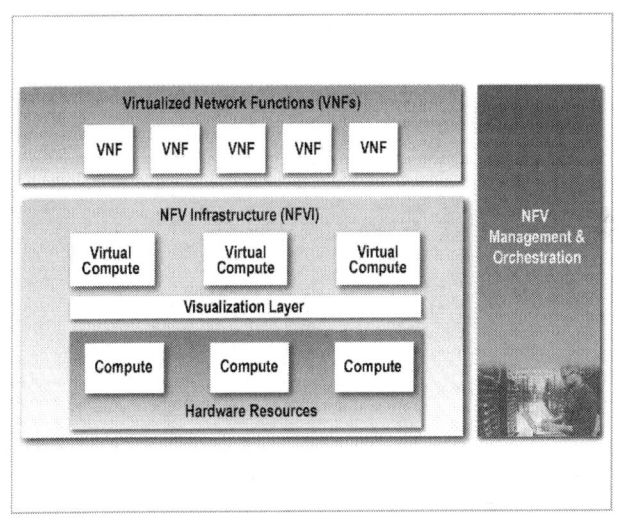

Figure 4.5. Network Function Virtualization (NFV) management and orchestration.

If NFV is used, the SDN can also act as a hypervisor for NFV virtual machines. Both technologies depend heavily on virtualization to enable network design and infrastructure to be abstracted in software and then implemented by underlying software across hardware platforms and devices. Although network virtualization segments differentiate virtual networks within one physical network or connects devices on different physical networks into one virtual network, SDN controls data packet routing through a centralized server.

Approaches for implementing cybersecurity protections include the following:

- Embed security within the virtualized network devices
- Embed security into the SDN servers, storage, and other computing devices

NFV establishes a virtualized networking environment dedicated to providing different network services. The security of data hosted in these environments largely depends on the degree of isolation between virtualized components. SDN is a new form of threat because the centralized software controller manages all network flows. While these issues are not 5G specific, its security framework needs to address these issues. Unlike an LTE network, which is owned by a single network operator, 5G networks are composed of multiple stakeholders providing specialized services (figure 4.6). Designing, building, deploying, and operating new transformative solutions securely will mandate effective and thorough due diligence across the entire solution ecosystem.

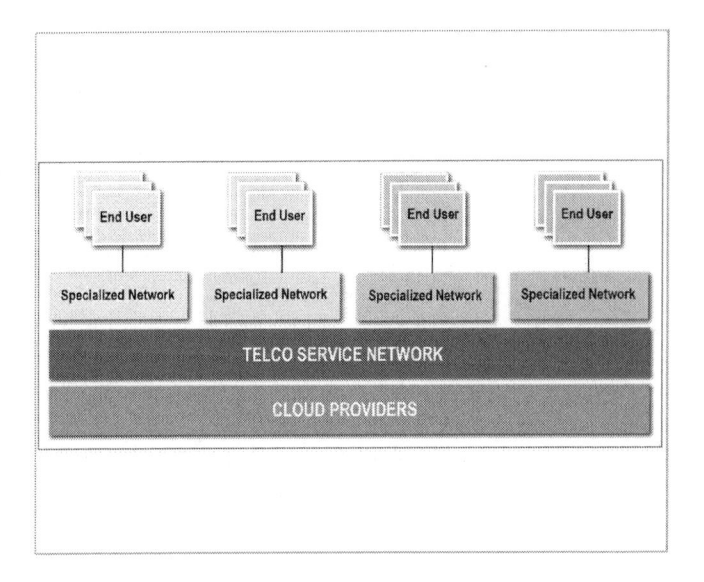

Figure 4.6. Multiple specialized networks constitute 5G networks.

These specialized networks may provide network services catered to the specialized end users. Traditional networks provide segment-based security by ensuring secure communication paths between the various parties. This will not be efficient in 5G environments that require end-to-end security to serve new forms of specialized networks. This will lead to a new trust model for 5G networks with an additional element of services compared to the traditional 4G trust model (figure 4.7).

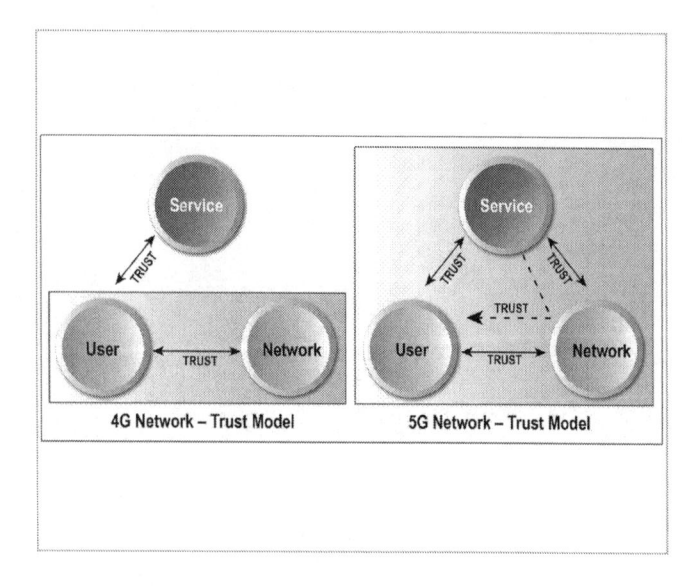

Figure 4.7. 4G and 5G network trust models.

With NFV, each network function no longer resides on its own proprietary hardware platform where this physical isolation provides a high level of protection. Instead, it now resides in software as a virtual network function (VNF) running on virtual machines (VMs) alongside other types of VNF, all sharing the same standard server hardware. Left unprotected, VNFs may interfere with one another, or there could be cross-contamination of malware from one VM or VNF to another, resulting in data leakage. In order to address this risk, some leading operators are already deploying containers.

Some popular IoT use cases assume the collection and storage of data at the edge of the network as well as the application of analytics to that data at the edge. This creates a scenario in which cost and space constraints may require that the VNFs and enterprise data and analytics software may share the same physical hardware. This requires a clear separation of duties between the enterprise and the network operator.

Zero Trust Security Model

The Zero Trust security model is centered on the belief that organizations should not trust anything inside or outside their perimeters. This model requires verification of anything and everything trying to connect to its systems before access is granted. This information security model abandons the castle-and-moat security model that focuses organizations on defending the system perimeter while assuming everything inside the barrier doesn't pose a threat. Security and technology experts criticize the castle-and-moat approach by pointing to the fact that some of the worst data breaches happened after hackers gained access inside corporate firewalls, at which point they were able to move through internal systems without resistance. With cloud, most of the data will be logically "outside" and in multiple systems in multiple locations, which makes the zero trust security model a necessity for the modern business. The Zero Trust approach uses existing technologies and governance processes in securing the enterprise IT environment. It calls for the use of micro-segmentation and more granular perimeter enforcement based on users, their location, and additional data to determine whether to trust an entity seeking access to an enterprise asset. Zero Trust uses multi-factor authentication, IAM, orchestration, analytics, encryption, scoring, and file system permissions. This puts the focus on user identity and resource access management regardless of the physical or logical location. It also uses governance policies that minimize access required to complete a specific task. When designing and deploying transformational solutions across enterprise, cloud and 5G networks, and MEC environment, the Zero Trust paradigm must be extended to include all associated software-defined networks (SDN).

As proposed by Huawei Technologies[10], a leading global 5G telecom equipment provider, the trust model for 5G service delivery is considerably more versatile than the previous 3G and 4G wireless systems. The service provider or network provided may authenticate the user. Double authentication can also be used if desired. The network provider could also authenticate the service provider directly. If the user authenticates with the network, a connection can be made automatically to the service. This improved approach should also increase efficiency.

When network slicing is used, they are configured differently based on the user type and device. This is done to ensure end-to-end (E2E) security. This requires the implementation of a standardized framework for the 5G network architecture as depicted in figure 4.8.

---

[10] https://www.huawei.com/minisite/5g/img/5G_Security_Whitepaper_en.pdf

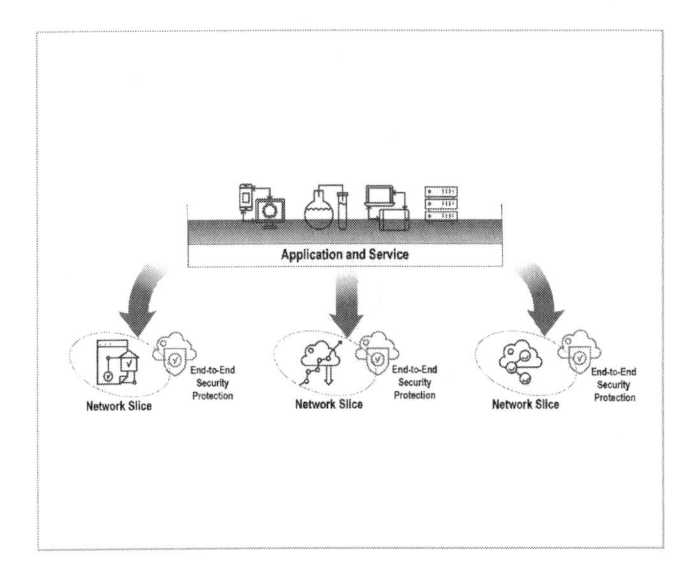

Figure 4.8. End-to-end security protection in 5G network architecture.

A *Harvard Business Review* study of hybrid strategies positions IT as an essential part of the business.[11] Enterprise composable infrastructures, continuous delivery release automation, and MEC can deliver increased control and the greater business benefits promised by 5G. By combining the impact of these three capabilities within a new Zero Trust network model, companies can successfully build and deploy more flexible, more efficient, and more secure ecosystems. This would enable these companies to effectively meet customer expectations, gain a competi-

---

[11] http://docplayer.net/44684280-Hybrid-it-takes-center-stage.html

tive advantage over competitors, and increase selling opportunities.

Successful Infrastructure Transformation

Few organizations have abandoned existing data center infrastructures to adopt a cloud-only approach. This is primarily due to their need to derive the maximum return from their in-place systems. Even most "born in the cloud" companies have processes and data that aren't appropriate for use in a public cloud. As a result, organizations today often operate mixed environments incorporating public and private clouds along with core systems of record like ERP. At the time of the Harvard study, 63 percent of those organizations were pursuing a "hybrid IT" approach. The most critical challenges associated with operating mixed infrastructure environments include the need for the following:

- New enterprise budgeting processes
- New technical baseline and procedures
- New IT service procurement processes
- Development of new products and services
- Establishing consensus around different organizational targets and goals
- Funding of training and education resources to acquire newly required skill sets
- Structural organization modifications

All these challenges require executive backing and organizational process changes. Key recommendations to those organizations seeking the benefits of digital transformation are to do the following:

- Rationalize automation tools and employee skill sets
- Adopt local composable infrastructure and continuous delivery release automation together
- Pair continuous delivery with composable infrastructure
- Identify and engage relevant cloud service and telecommunications providers
- Merge architects, developers, and operations into integrated product teams
- Build teams focused on individual products and their releases, crossing disciplines of design, development, and operations

Summary

Our current population has the opportunity to see the world transition from one age to the next. We should consider ourselves to be extremely fortunate, especially if you zoom out and take a distanced look at the history of humankind. If you lived in the Stone Age, you died in the Stone Age. The generation before you and after you also lived and died in the Stone Age. If you lived in the Agricultural Age, you died in the Agricultural Age. These eras lasted centuries longer than that of the Industrial Age, which still housed three generations.

Compared to the masses of people who have come and gone, only a few have been able to see the world transition from one age to the next. And what are the driving forces behind these rapid global transitions? Technology and innovation, something that we can access easily and effortlessly in our day-to-day life.

There are plenty of people alive today who saw the transition from the Industrial Age to the Information Age. The 1950s and 1960s marked a significant shift in focus to information technology. But unlike the Industrial Age and those before it, the Information Age will not drag on for multiple generations. Entire generations have the opportunity to watch the world transition from the Information Age to the Virtual Age.

Cloud computing is compounding this accelerating rate of change. Physical labor or presence is no longer a requirement for change. Intimate knowledge of new technology is no longer relevant. We are now able to collaborate on a global scale with a focus on relationships and results. We can zoom out from individual technology and instead see the impact that it will have on the world. Increased access has given more people a fighting chance to share their solutions, increasing competition, and the push for speedy results.

In my lifetime, I have been able to physically travel around the world at a speed that was thought to be impossible a hundred years ago. More importantly, I have been able to build a business that reaches people around the world in real time. The speed and software available to us now were merely dreams at the dawn of the Information Age.

Think of all the people that you can call, message, or see on your computer screen within seconds. All these people provide the opportunity for collaboration, innovation, and disruption. It took less than a year and a handful of connections to make Source-Connecte a reality. In less than a year, it supported and transformed an entire industry. (I'll discuss that more in Chapter 7.) I am immensely excited to see the transformations made possible within this new era.

5

Transformation Network

Chapter 4 described composable infrastructures and how transformational enterprises must now fold cloud and edge computing capabilities into their product and services catalog. The Achilles heel of any transformative business model, however, is their reliance on ever increasing amounts of data that need to be transported quickly across wide area networks and processed at edge computing end points. To meet this expected demand, the global telecommunications industry is rapidly moving toward a future in which networks must have the agility, flexibility, and scalability to deliver aggregated capabilities through fully programmable networks. This capability delivers software-defined service domains that can seamlessly interact with other independent service domains in a multitenant environment. This is the transformational 5G network that will result in a telecommunications industry capable of evolving and adapting to the emerging networked society requirements.

Historically, sovereign nations have managed their telecommunications networks as national assets. The political negotiations that drove that history led to underlying technological choices and today's heated international competition around 5G network deployments. In the early days, telecommunications technology was dominated by American and European players. The rapid economic rise of China, however, has now made that country a formidable player in the international negotiations that will dictate the next generation of global telecommunications rules. As the world's second-largest economy, the Chinese market is expected to represent more than half of the 1.3 billion global subscribers to 5G networks by 2023. In comparison, the U. S. and Europe combined are expected to sign up just 337 million 5G subscribers. The Chinese telecommunications giant Huawei is now the world's largest telecoms equipment manufacturer. Western nations fear that Huawei's dominance of 5G technology could give the Chinese government backdoor access to Western mobile networks and the application that may run over them.[12] This inter-

---

[12] https://www.politico.eu/article/5g-telecommunications-infrastructure-china-us-eu-qualcomm-nokia-ericsson-huawei/

national competition will determine the availability of specific technologies and telecommunications resources in each geographic region. This will, in turn, determine the profitability of any technology-based product or service in those respective geographic marketplaces.

Chapter 5 provides a wireless network historical background and a snapshot of the 5G current marketplace. It will also provide a short primer on wireless networking technologies essential to delivering the products and services transformative enterprises need to deploy over the next few years.

Network History

Mobile wireless technology has evolved through multiple technology generations. Since the late 1970s, new generations of technology and wireless standards have been introduced every decade through the current transition between 4G and 5G capabilities. Each generation has increased its value to society exponentially as each has enabled global technology advancements. While existing generations use the low to mid-band spectrum (less than 6GHz, or sub-6), 5G uses millimeter-wave (mmWave) spectrum also.

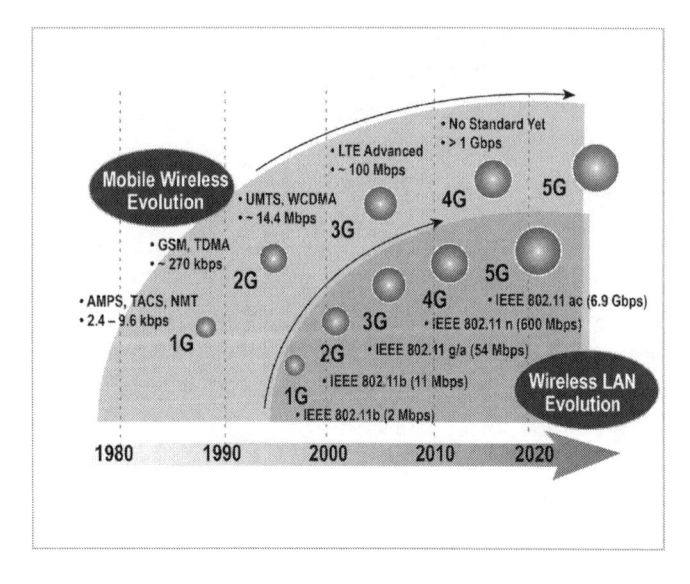

Figure 5.1. Mobile wireless evolution and wireless LAN evolution[13].

Voice Calls (1G)

The first generation (1G) was introduced in the late 1970s and fielded in the early 1980s. There was limited emphasis on data transfer capability (~2.4 Kbps), and these networks used analog signals to transfer cell users between a network of distributed base stations on cell towers. This generation used standards like AMPS (advanced mobile phone system developed by AT&T for the U. S.) and TACS (a variant of advanced mobile phone system) which was chosen

---

[13] Source: https://www.researchgate.net/figure/Wireless-technology-evolution_fig1_322584266

by the first two UK national cellular systems in February 1983.

Messaging (2G)

2G mobile networks provided the first digitally encrypted telecommunications in the 1990s. This technology improved voice quality, data security, and data capacity. Limited data capability was provided using circuit-switching under the European Telecommunications Standards Institute (ETSI) Global System for Mobile Communications (GSM) standard. Improved data rates were brought to the market in the late 1990s by using 2.5G and 2.75G technology, which were named GPRS (general packet radio service) and EDGE (enhanced data rates for GSM Evolution). This provided improved data transmission rates as a backward-compatible extension of GSM. IT also introduced data transmission via packet switching and served as a stepping-stone to 3G technology.

Limited Data: Multimedia, Text, Internet (3G)

The late 1990s and early 2000s brought 3G networks. These systems fully transitioned to data packet switching, which delivered faster data transfer rates. This enabled data streaming, mobile internet access, fixed wireless access, and video calls. 3G networks now have data speeds of 350 Kbps when mobile and can reach 1 gbps when stationary using standards such as UMTS (universal mobile telecommunications system) which is based on the GSM standard. This standard is developed and maintained by the 3GPP (3rd Generation Partnership Project).

True Data: Dynamic Information Access, Variable Devices (4G and LTE)

4G network services, introduced in 2008, had speeds that were ten times faster than 3G. This was done by leveraging all-IP networks and relying entirely on packet switching. These networks enhanced the quality of video data due to larger bandwidths allowing for increased network speed. The introduction of the LTE network later set the standard for high-speed wireless communications on mobile devices and data terminals. The current version of LTE can support ~300 Mbps.

5G

The precise capabilities and extent of adoption of 5G are still being finalized. Data transfer speed, volume, and latency depend on the spectrum bands used and the network usage context (fixed or mobile).

MmWave spectrum is a high-frequency technology that lies between 30 GHz and 300 GHz. It is attractive because its shorter wavelengths create narrower beams, which provides better resolution and security for data transmission. This option can carry large amounts of data at increased speeds with minimal latency. Another important point favoring this technology is the amount of mmWave bandwidth available. This availability improves data transfer speed and avoids the congestion that exists in lower spectrum bands. A 5G mmWave system requires a significant infrastructure build but could reap the benefits of data transferred at up to twenty times the speed of current 4G LTE networks. Components for mmWave are also smaller than those for other bands, which allows for more compact deployment on wireless devices.

Sub-6 (spectrum below 6 GHz) can provide broad area network coverage with a lower risk of an interruption than experienced with mmWave. This is a characteristic of its longer wavelength and greater capacity to penetrate obstacles. Deployment of this technology requires less capital expenditure by network operators and fewer base stations. Such a system would also be able to use the existing 4G infrastructure.

A mmWave 5G network, for example, could enable incredibly fast speed for fixed local area networks under conditions that did not limit wave propagation but would not be able to maintain those high speeds at extended range. A sub-6 5G network may have a lower maximum speed than mmWave but conversely could cover a much broader area at a higher quality service level. Use case and business model completion across the IT marketplace will ultimately determine global 5G standards.

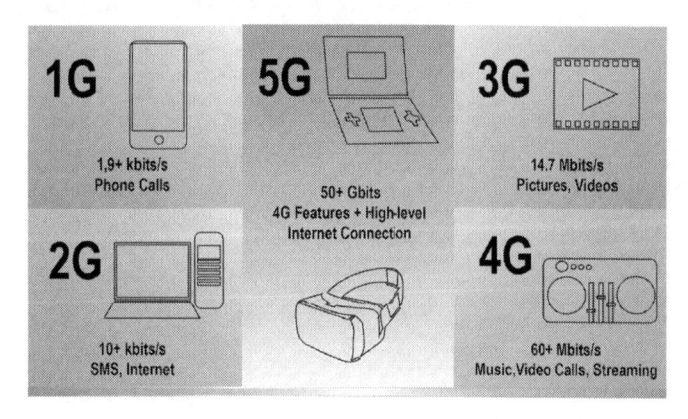

Figure 5.2. The evolution of wireless generations..

International 5G Network Competition

A country's choice of whether to standardize its 5G network on mmWave or sub6 will drive the product design and manufacturing for that country's 5G supply chain. Since 5G enables transformative business models, this choice will have an outsized effect on commercial development and growth across essentially every industry in every country.

Wireless technology generation transitions before 5G also had global commercial, competitive, and security implications. Germany gained the first competitive advantage in 2G. Being in Europe, this enabled companies like Nokia and Ericsson to roll out more advanced devices earlier than their competitors. They were already transitioning to 3G in the 2000s when the United States was still implementing 2G. Europe lost the lead during the 3G transition due to regulatory 3G spectrum auctions, at which time Japan took the global lead in the 3G race. The United States caught up to Japan after a long rollout, but the lag delivered a huge blow to US businesses as Japan quickly advanced 3G business models. Many US wireless technology companies failed or were absorbed into foreign companies during this period.

The late US surge in 3G investment gave it a head start during the 4G and 4G LTE transition. The FCC also opened licenses for more bandwidth and set regulations that promote the rapid expansion of the 4G networks. However, the Japanese industry didn't move quickly enough to develop the 4G ecosystem, and consequently fell behind. This enabled the United States to take the early lead in smart devices, displacing Japanese operating systems globally.

AT&T and Verizon rapidly deployed LTE in the early 2010s. Using 700 MHz spectrum they won at the 2008 auction. Finland and the United States were the first two countries to deploy comprehensive LTE networks. This technology delivered approximately ten times network performance of previous 3G networks. This huge improvement drove rapid new handsets adoption. The new devices moved much more data and were computationally much faster. US companies (e.g., Apple, Google, Facebook, Amazon, Netflix) built new applications and services by taking advantage of wider bandwidth and the new handset capabilities. As LTE deployment spread in other countries, those handsets and applications spread globally and drove US dominance in wireless and internet services.

5G International Marketplace

National 5G capability can be compared across spectrum availability, 5G trials, national regulator 5G road maps, government commitment, and industry. Spectrum availability is the most important factor since many of the others are dependent on that availability. The ongoing marketplace debate is around the allocation of sub-6 versus mmWave spectrum. In the United States, there is additional concern around allocating or sharing a government-owned spectrum with the commercial sector. Telecommunications carriers building infrastructure can either leverage existing 3G and 4G infrastructure as a stepping stone to get to full 5G capability or make a large up-front investment to build out new 5G network infrastructure. These decisions will also determine the most profitable transformative business models.

	1GHz	3GHz	4GHz	5GHz	24-28GHz	37-40GHz	64-71GHz
					24.25-24.48GHz	37-37.6GHz	
🇺🇸	600MHz (2x3.5MHz)	2.5GHz (LTE B41)	3.55-3.7GHz, 3.7-4GHz	5.9-7.1GHz	24.75-25.25GHz, 27.28-38GHz	37.5-40GHz, 47.5-42.2GHz	64-71GHz
🇨🇦	600MHz (2x3.5MHz)				27.5-28.35GHz	37.5-40GHz, 47.5-48.2GHz	64-71GHz
🇪🇺	700MHz (2x30MHz)		3.4-3.8GHz	5.9-8.4GHz	24.5-27.5GHz		
🇬🇧	700MHz (2x30MHz)		3.4-3.8GHz		26GHz		
	700MHz (2x30MHz)		3.4-3.8GHz		26GHz		
🇮🇹	700MHz (2x30MHz)		3.46-3.8GHz		26GHz		
	700MHz (2x30MHz)		3.6-3.8GHz		26.5-27GHz		
🇨🇳			3.3-3.6GHz	4.8-4.9GHz	24.5-27.5GHz	37.5-42.5GHz	
🇰🇷			3.4-3.7GHz		26.5-29.5GHz		
🇯🇵			3.4-4.2GHz	4.4-4.9GHz	27.5-29.5GHz		
			3.4-3.7GHz		24.25-27.5GHz	36GHz	

New 5G Band
- Licensed
- Unlicensed/Shared
- Existing Band

Figure 5.3. Global bands of the 5G spectrum[14].

OPERATOR	COUNTRY	DETAILS
Megafon	Russia	Trial 5G networks for eleven cities hosting the Fifa 2018 World Cup. Trial involving M2M and human users.
KT, SKT, LG	South Korea	The Ministry of Science, ICT and Future Planning plans to invest up to KRW 68.2 billion to ensure the roll out of a fully commercial 5G service by 2020. Trial services are planned for 2017, with launch of some limited demonstrators services at the Pyeong Chang 2018 Winter Olympics
NTT DoCoMo	Japan	Field trials underway. Commercial launch of 5G for summer Olympic Games in 2020 as first phase (serving stadia and other areas); second phase focused on latency and higher frequencies to follow, potential in 2022/2023 time frame
Orange	France	Trials in Belfort to end 2016
Softbank	Japan	5G trials in Tokyo
Verizon	United States	Lab trials underway. Technology field trials to begin in 2016. Possible 5G commercial launch within 2017
Deutsche Telekom	Germany	5G:haus innovation lab launched in March 2015, working with multiple partners
Telstra	Australia	Field trials, proofs of concept and radio test bed initiated in March 2015
América Movil	Brazil	Test bed announced in October 2015

Figure 5.4. Operator table - 5G trials and commitments

---

[14] Source: https://www.researchgate.net/figure/The-global-snapshot-of-5G-spectrum_fig1_338870949

The 5G transition will carry ever more significant business risks and rewards than the earlier generational shifts. The global 5G leader will make hundreds of billions of dollars in revenue over the next ten years while simultaneously enjoying widespread job creation in the wireless technology sector. 5G can also revolutionize other industries through autonomous vehicles, the Internet of things (IoT), and the replacement of current fiber-optic household network backbones. The country that leads will own many of these innovations and set global standards.

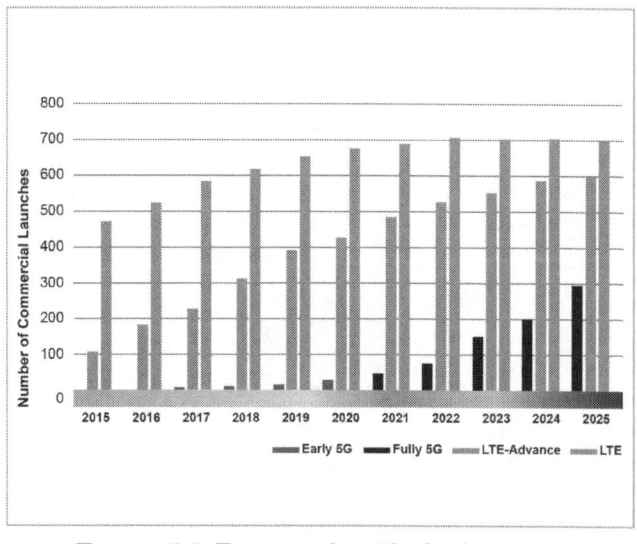

Figure 5.5. Forecast for 5G deployment.

5G Network Enhancements

The International Telecommunication Union (ITU) has set forth aggressive requirements on network speed and responsiveness, in addition to faster data transfer and the ability to handle more data across many more devices. These requirements are substantially superior to previous wireless technology generations and have set off the rapid development of many new wireless network capabilities. Companies need to be familiar with the overarching technologies so that they can determine what type of network is needed for their intended innovations or applications.

Beamforming

Beamforming directs radio energy through the radio channel toward a specific receiver (top left quadrant of figure 5.4). By modifying the transmitted signal phase and amplitude, receiver signal strength, and end-user throughput are increased. When receiving, this capability can collect signal energy from a specific transmitter. Multiple input and multiple output (MIMO) is the grouping of multiple antennae at the transmitter and the receiver to send more information over the spectrum. Massive MIMO and beamforming are forms of MIMO that focus energy from a large number of antennae on a very narrow beam of signal that facilitates the handling of a myriad devices within an area. These technologies also support the use of very short wavelengths by countering path loss or the loss of energy in the radio wave as it travels through space.

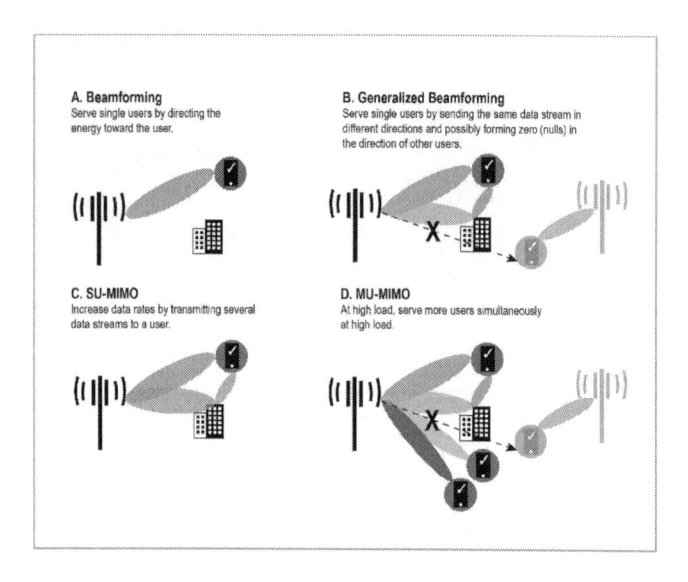

Figure 5.6. Beamforming and MIMO[15].

MIMO

Multiplexing is when a radio separates the uplink and downlink transmissions using separate spectrum resources to better share and optimize the resource. MIMO (multiple input, multiple output) refers to wireless network spatial multiplexing techniques. The term describes the transmission of multiple data streams, with the same time and frequency resource. In this approach, each data stream can be beamformed. MIMO increases throughput by using high-quality signals to receive multiple data streams

---

[15] Source: https://techblog.comsoc.org/2020/07/16/tutorial-on-advanced-antenna-systems-aas-for-5g-networks/

at a reduced power per stream. MIMO works in both uplink and downlink. Single-user MIMO (SU-MIMO) transmits one or multiple data streams, called layers, from one transmitting array to a single user whereas multiuser MIMO (MU-MIMO) sends different layers in separate beams to different users using the same time and frequency resource, which increases network capacity.

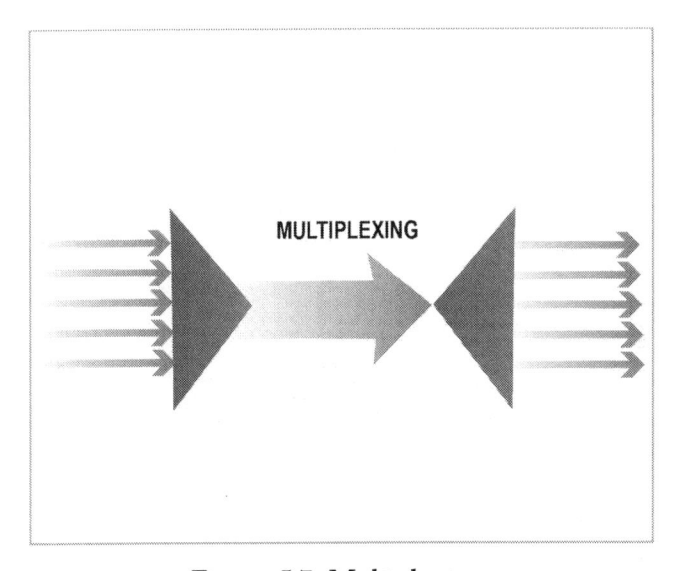

Figure 5.7. Multiplexing.

Standard MIMO networks use two or four antennas. Massive MIMO, however, uses a very high number of antennas. There's no standard number that constitutes massive MIMO; the term is used to describe tens or even hundreds of antennas. Huawei, ZTE, and Facebook, for example, have demonstrated massive MIMO systems with 96 to 128 antennas. It also makes the network far more resistant to interference and intentional jamming. Massive MIMO can multiply the capacity of a wireless connection without requiring more spectrum, which could potentially deliver a fifty-fold increase in the future. A massive MIMO network is also more responsive to higher frequency band devices, which improves coverage. Massive MIMO with beamforming enhances end-user experience by increasing transmission effectiveness. A stronger signal, higher data throughput, and greater transmission distances increase capacity in congested cities.

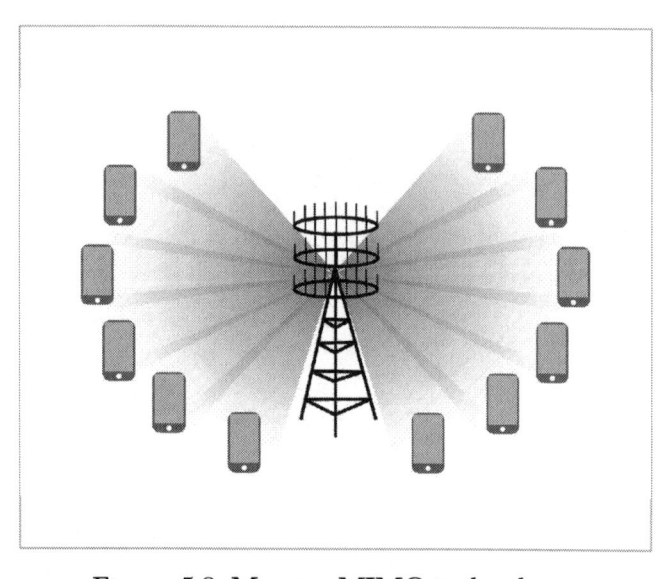

Figure 5.8. Massive MIMO technology.

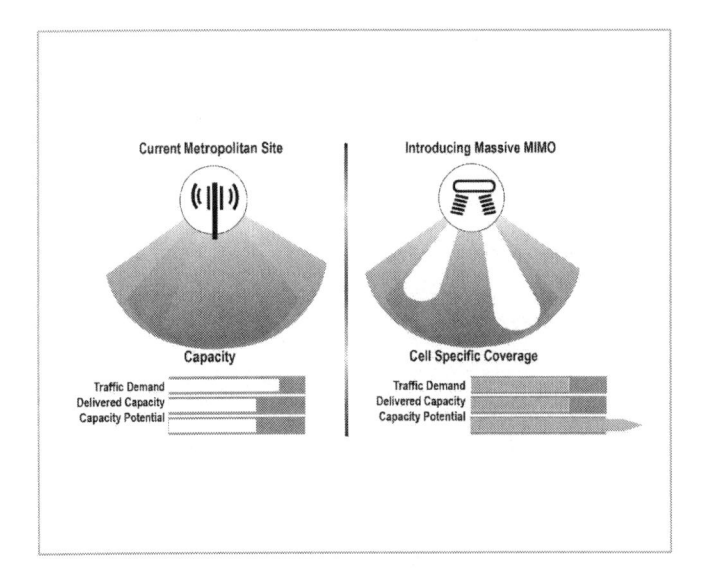

Figure 5.9. Current metropolitan site vs. massive MIMO[16].

Advance Antenna Systems

An advanced antenna system (AAS) combines an AAS radio and additional features that include beam-forming and MIMO. This solution provides much greater adaptivity and steerability when compared to the conventional antenna, when changing antenna radiation patterns, rapidly time-varying traffic, and multipath radio propagation are a concern. Addition ally, multiple signals can be received or transmitted simultaneously with different radiation patterns. The beams formed by an AAS are constantly adapted to

---

[16] https://www.5gworldpro.com/5g-knowledge/what-is-massive-mimo-mmimo-in-5g.html

the surroundings to give high performance in both uplink and downlink.

Small Cells

The use of additional cells in a given area is so that there are fewer communication devices with the base station over a given area. Small cells are used with the mmWave system and can potentially transmit more data and, when used with network densification, can reduce path loss.

Device-to-Device Communications

To reduce latency, device-to-device communications coupled with edge computing will reduce communications response times to a level that will support analysis and response for time-critical use cases across autonomous vehicles.

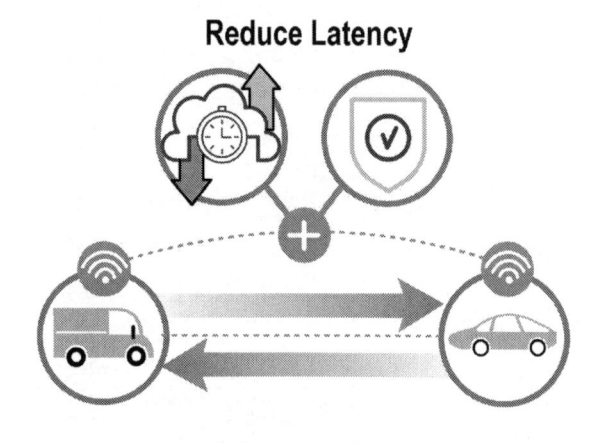

Figure 5.10. Reduce latency.

Network Slicing

Using network slicing, telecommunications operators can assign specific network slices on a single physical network to specific devices and applications. This is a virtual networking architecture that creates multiple virtual networks atop a shared physical infrastructure. Physical components are secondary and logical partitions are paramount. With this approach, capacity can be dynamically assigned based on need. When used with shared resources, network slicing can provision slices that are devoted to logical, self-contained, and partitioned network functions. Using network slicing, network operators can deploy only the functions needed to support specific customers within specific market segments.

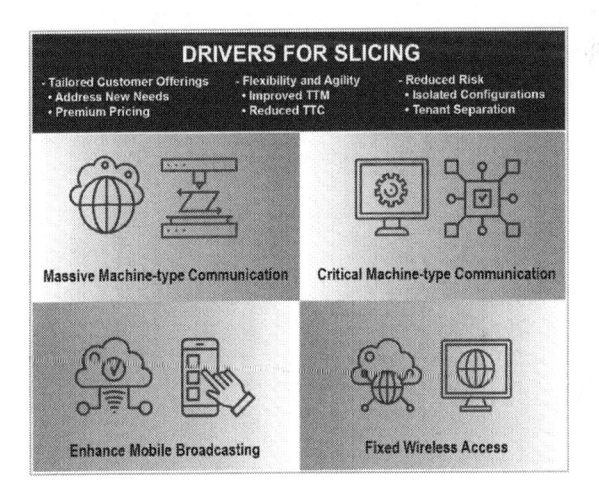

Figure 5.11. Main drivers behind network slicing[17].

---

[17] Source: https://www.ericsson.com/en/blog/2019/5/

Distributed Cloud

A distributed cloud maintains cloud computing flexibility while simultaneously hiding the complexity of the network infrastructure. Application components are placed in an optimal location to use compute and data storage services of the distributed cloud. This, in essence, is the application of cloud computing technologies to interconnect data and applications served from multiple geographic locations. This approach can be used to distribute data for use by distributed applications, data protection, and compliance with data sovereignty laws. The distributed cloud approach increases capacity, availability, and coverage while also limiting data transfer requirements. The placement of application components at edges will depend on the behavior of the application and the available infrastructure resources. A distributed cloud solution enables edge computing through the use of micro and small data centers. Localized data collected using local and wide area networks are stored in the central cloud and integrated on the edge computing architecture. Micro and small data center locations may be dependent on the CSP's network topology and use cases requirements.

The distributed cloud is reliant on efficient management and orchestration that enables automated application deployment across heterogeneous clouds. It enables service creation and instantiation

---

highlights-of-key-end-to-end-network-slicing-capabilities

in environments provided by multiple partners and suppliers. Distributed cloud management solutions also offer discovery, onboarding, and auto enrollment of edge components. Applications deployed in the distributed cloud present their capabilities through service publishing and discovery. The cloud infrastructure and connectivity layers expose their respective capabilities through APIs. Application developers will use these APIs for mobile connectivity.

Summary

The evolution of telecommunications reflects the evolution of technology as described in Chapter 4. Unlike the existence of technology, the journey from 1G to 5G has taken place in less than a hundred years. Our current existence in the new Virtual Age, along with the rise of 5G, both argue for and prove the existence of globalization. Don't believe me? Let's put aside telecommunications for a moment and reflect on COVID-19.

COVID-19 is the first pandemic to rock the world since 1918. This time, information (and misinformation) about the pandemic spreads faster than the pandemic itself. Within seconds, we can connect with friends, exchange ideas with colleagues, and hear from thought leaders experiencing COVID thousands of miles away. Although the world has spent most of the pandemic existing in physical isolation, we are still connected in ways that we could not connect before. The virus transcends borders. COVID-19 is a truly global experience.

People holding isolationist ideology may resist the idea of a global society, but they can no longer deny its existence. Globalization has existed for decades. Failing to embrace and innovate with a globalist mindset traps companies and nations in the Industrial Age.

The world is rapidly accelerating toward a fully realized Virtual Age. Understanding 5G, and collaborating under a globalized 5G network, is at the cornerstone of that acceleration. Before we operate under this network, we have some negotiations to settle. We are currently in the throes of a new cold war. Ideology has simply been replaced with control of the economy. The world has quickly merged its localized telecommunications networks into a larger network spanning the entire planet. Leaders ask, "Who will control it?" If the evolution of cloud computing has taught us anything, control is not the answer—collaboration is.

Innovators, world leaders, and individuals must embrace a globalized 5G network. It benefits us all. Standards and innovations will shape the framework and standards under which 5G operates. The capabilities and usage of 5G will expand or limit how we collaborate, build, and solve the world's largest problems. Failing to accept this evolution on a global scale only denies the current abilities to communicate and conduct business.

6

Transformation Innovation

As visualized by the business models in Chapter 1, successful digital transformation means effectively deploying business models that collect, analyse, and transform data into information that's capable of delivering value to an organization's customers.

This entails the integration of digital technology into every business process. The accomplishment of this goal results in fundamental changes to not only how businesses deliver value but also, in many instances, changes the essence of the value being delivered. Born-in-the-cloud companies often achieve this by eliminating the physical components of an old-business model and replacing them with provisioning data or with information exchange. Uber removed physical vehicles from the taxicab model by replacing it with an information exchange service between those that wanted transportation and those that were willing to offer a ride. Airbnb did the same thing by replacing physical hotel building with an information exchange service between travelers and homeowners.

Cloud services and capabilities are essential to digital transformation initiatives because cloud computing decouples data and information from its underlying infrastructure. This process is more involved than just using someone else's server and dragging some folders from your desktop to a shared cloud drive. This is why digital transformation can be so overwhelming.

When starting this journey, companies must first reeducate themselves on the data they collect, manage, and deliver. Data lies at the core of all business models, and in today's mobile and globally connected business environment, this critical business component can be moved, viewed, manipulated, and destroyed at will unless proper safeguards are put in place. This reeducation starts with coming to grips with understanding the data lifecycle and how it interacts with the organization's business or mission model.

Data Security

Data security is centered on three concepts:

- *Confidentiality*: Preventing unauthorize people from accessing sensitive data.
- *Integrity*: Maintaining the consistency, accuracy, and trustworthiness of data over its entire life cycle.
- *Availability*: Ensuring that those that are authorized to access data can access it when the need occurs.

Often referred to as the CIA Triad, the achievement of these is the goal of every information security professional. Data security must be enforced throughout the six phases of the data lifecycle, which are:

- *Create.* The acquisition or generation of new digital content, or the alteration/updating of existing content. This phase is the preferred time to classify content according to its sensitivity and value to the organization.
- *Store.* Putting data into a storage repository which occurs nearly simultaneously with creation. Initially, storage is temporary until the data is used for some purpose. The repository is often a mobile device or a cloud service. Legal, regulatory, privacy, and national data sovereignty rules can often dictate where data is stored. Privacy concerns also need to be addressed in this and all subsequent phases as well.
- *Use.* Viewing or processing, or otherwise used in some activity, not including modification. Legal, regulatory, and privacy guidelines may restrict available data access functions or business activities based on user role, access device, and/or physical location. Data is most vulnerable during this phase because it might be transported to unsecured locations such as workstations. The use of data loss prevention (DLP) and data right management (DRM) technologies in this phase to enforce organi-

zational data governance policies are crucial in this phase. Failure to establish appropriate data governance policies or a failure to invest in DLP and DRM for enforcement of those policies can result in data loss or breach along with grave reputational damage and large legal costs.

- *Share.* Information made accessible to others during this phase. This phase also requires the effective use of DLP and DRM technologies. Since shared data is no longer under the organization's control, the installation of DRM software agents onto any device that may have access to organizational data is mandatory if persistent data protection or continuous data audit is needed. DLP is used to detect unauthorized sharing while DRM is used to maintain control over the data.
- *Archive.* In this phase, data are leaving active use and entering long-term storage. Data placed in archives must still be protected according to their classification. While cost-versus-availability decisions may affect data access procedures, regulatory legal and privacy requirements must continue to be honored.
- *Destroy.* The permanent destruction of data using physical or digital means is done in this last phase. The only practical means to destroy data in a cloud computing environment is through crypto-shredding. This is only pos-

sible if the data has been continuously encrypted using keys created and maintained solely by the organization.

Across these six phases, the organization's business or mission model will dictate data access requirements and authorization guidelines for the following:

- Multiple organizational actors, both human and nonhuman
- Actors playing multiple roles within the business process or function
- Role responsibilities and functions being executed on multiple and differing devices
- Multiple devices, possibly from individual actors playing multiple roles, accessing data and business processes from varying global locations
- The possible channels (i.e., WiFi, virtual private network (VPN), wireline, broadband wireless, etc.)

This scenario is neither rare nor represents a small subset of the business. It is the environment that *all* organizations that wish to infuse digital technology into all areas of their organization *must* embrace. In other words, doing this represents table stakes for digital transformation.

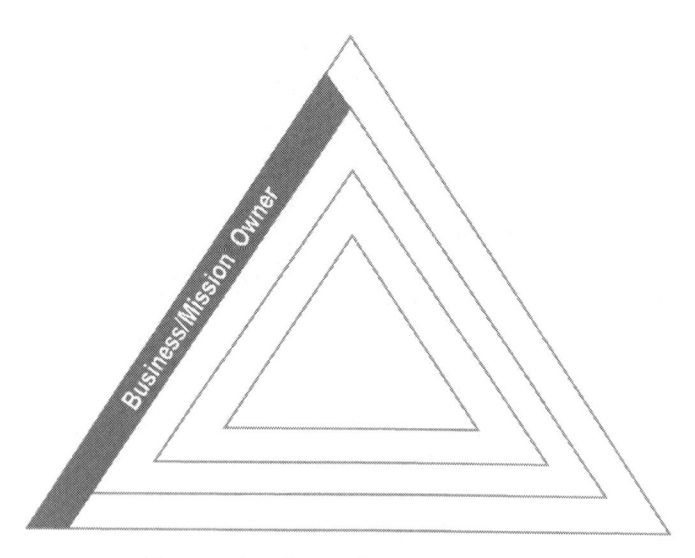

Figure 6.1. Digital transformation triangle - first leg.

Once a business or mission model is documented (i.e., API management, big data analytics, e-commerce, etc.), the organization must then identify the data that feeds these process, which role executes each process, which organizational actors can fulfill each role, and a data and information use framework to address access from every authorized device, channel, and location option. Accomplishing this task falls within the purview of the business/mission owner and represents the most important leg of the digital transformation triangle.

Risk Management

Hybrid IT is multifaceted and involves multiple participants, all of whom operated within a shared risk model. Enterprises that use cloud computing take risk into account as a component of an interrelated ecosystem that may not be controlled by the organization's IT department. Risk-management design is traditionally targeted for low-uncertainty environments that have few interconnections. The risk in the modern networked world contains high uncertainty within dynamically changing, interconnected systems. Key cloud computing risks include the following:

- Inability to fulfill financial goals
- Failure to remain within the context of corporate organization and culture
- Unsurmountable difficulties in integrating the cloud services involved
- Inability to comply with legal, contractual, and moral obligations
- Inability to recover from a disaster
- Technically inadequate cloud service
- Inadequate solution quality

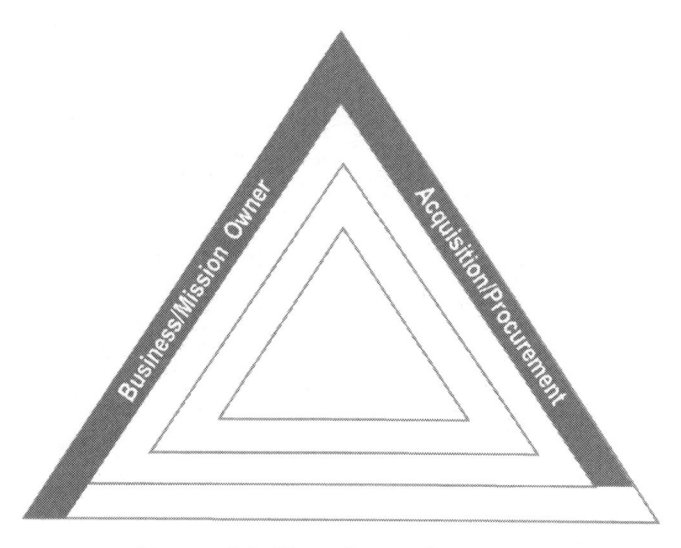

Figure 6.2. Digital transformation triangle - second leg.

Assessing Risk

Risk assessment should evaluate relevant financial, culture, service integration, regulatory compliance, business continuity, and business or mission system quality factors. Financial risk impact is always critical as it directly drives the return on all investments associated with a cloud computing transition. Since cloud service consumption costs are directly related to workload and revenue, risk assessment is a seminal component for determining cloud migration return on investment and the probability of attaining all related ROI (return on interest) targets. Probabilities of attaining ROI targets are often central to conversations with the organization's acquisition and procurement teams, which brings in a second triangle leg.

Modifying these probabilities and the associated cost of mitigating risks are oftentimes met by budgetary constraints. The critical assessment factors for cloud ROI risk probability are the following:

- Infrastructure utilization
- Speed of migration to cloud
- Ability to scale business/mission processes
- Quality delivered by the new cloud-based process

These four factors directly drive digital transformation ROI because they affect revenue, cost, and the time required to realize any investment return. Differences between actual and projected values indicate a likely failure to achieve the desired goals.

Although business alignment is always a primary digital transformation driver, ROI remains a key decision component. This metric should, however, be addressed from multiple vantage points to include the following:

- Workload versus utilization
- Workload type allocations
- Virtual hardware instance to physical asset ratio
- Speed of operational cost reductions
- Optimizing cost of capacity
- Optimizing total cost of ownership
- Process time reductions
- Product quality improvements
- Customer experience enhancements

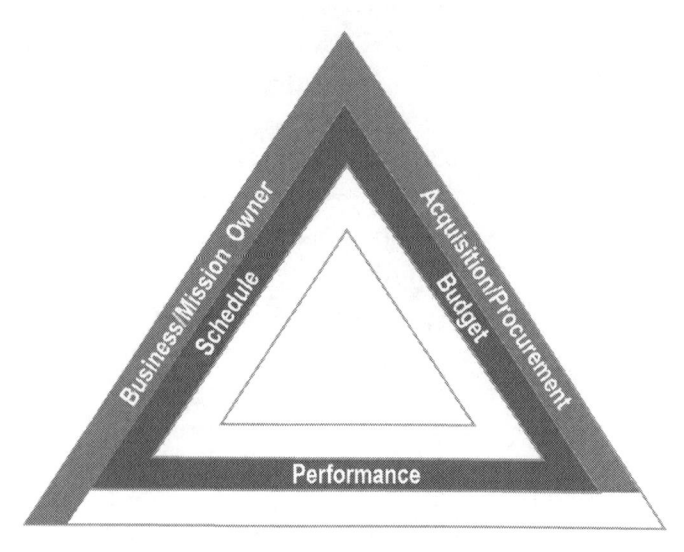

Figure 6.3. Digital transformation triangle - first interior legs.

Mitigate Risks

Decisions on which business/mission model risks require mitigation are implemented through the use of security controls. A control will restrict a list of possible actions down to what is allowed or permitted by the organization. Encryption, for example, can be used to restrict the unauthorized use of data. This is why an understanding of data function, role, actor, device, channel, and location is so important. Once these factors have been documented and understood, the appropriate controls can be designed and applied to the business/mission process in order to safeguard data and control access to it. These controls can be preventative, detective (monitoring), or corrective in nature. IT governance is, in fact, the selection and application of security controls that adequately protect organizational data while simultaneously minimizing operational friction or disruption. The security control continuum extends over three categories:

- Management (administrative) controls: policies, standards, processes, procedures, and guidelines set by corporate administrative entities (i.e., executive to mid-level management)
- Operational (and physical) controls: operational security (execution of policies, standards and process, education, and awareness) and physical security (facility or infrastructure protection)
- Technical (logical) controls: Access controls, identification and authentication, authorization, confidentiality, integrity, availability, and non-repudiation

They also encompass the following types:

- Directive controls: often referred to as administrative controls, advise employees of the behavior expected of them during their interfaces with or use of information systems
- Preventive controls: include physical, administrative, and technical measures that preclude actions that violate policy or increase the risk to system resources

- Deterrent controls: use warnings and a description of related consequences to prevent security violations
- Compensating controls: Also called an alternative control, a mechanism that is put in place to address security requirements deemed impractical to implement
- Detective controls: Refer to the use of practices, processes, and tools that identify and possibly react to security violations
- Corrective controls: involves physical, administrative, and technical measures designed to react to a security-related incident in order to minimize the opportunity for an unwanted event to reoccur
- Recovery controls: restore the system or operation to a normal operating state once integrity or availability is compromised

The costs associated with the implementation of any control should be weighed against organizational business/mission goals. The outcome of the ensuing risk-reward debate normally results in digital transformation project budget decisions and business model performance options. Also, as all project managers know, these decisions tend to have major effects on an attainable digital transformation schedule.

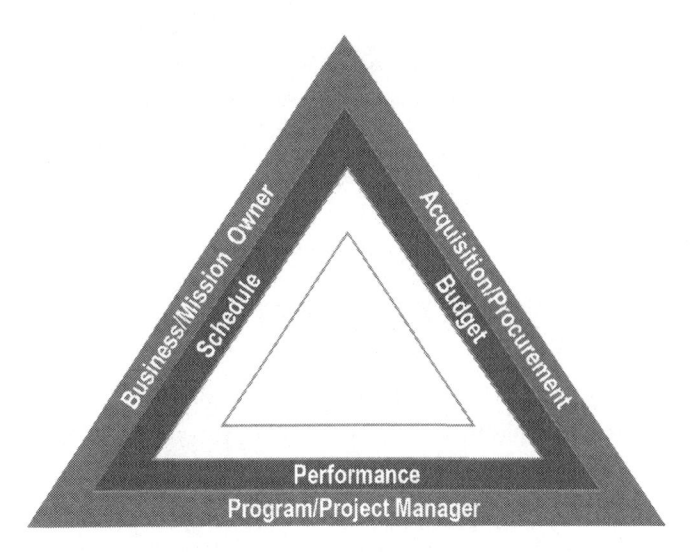

Figure 6.4. Digital transformation triangle - first and second layers complete.

Monitoring Risk

With risk mitigation decision now in place, the program/project manager is put in place to ensure that the levels of digital transformation risk continue to be acceptable. Risk management doesn't end here but must be repeated at every significant digital transformation decision stage. Since cloud service procurement is an operational expenditure, the digital transformation procurement management process must include a continuous service monitoring and verification component. Cloud service risk should be re-evaluated if the service is modified or if new service options are made available in the broader marketplace. This requirement is the basis for maintaining and updating industry benchmarks for every critical business process service.

5G Business Models Scenarios

There are three primary scenarios for defining how different parties interact over a 5G network infrastructure. Based on the key 5G capabilities, they are enhanced mobile broadband (eMBB), ultrareliable and low-latency communications (uRLLC), and massive machine type communications (mMTC).

Enhanced Mobile Broadband

The key capability for enhanced mobile broadband (eMBB) is speed. This advantage is primarily exploited for use with consumer-facing 5G wireless communications services. The eMBB interface uses massive MIMO and the mmWave spectrum, along with other technologies, to provide three key benefits: (1) it delivers a faster speed of transmission, (2) increases the capacity to handle greater volume of data per area, (3) and lowers the cost of data transfer over the network.

This expands service quality for the user and improves the download capacity that service providers can give to the network. eMBB is used to wirelessly replicate the experience from a fibre-optic cable network on mobile devices. Relevant applications include data-intensive services within densely populated areas, such as virtual or augmented reality, live-multiplayer gaming events, or mobile video-streaming services.

Ultrareliable and Low-Latency Communications

Ultrareliable and low-latency communications (uRLLC) depend on the speed of transmission. However, in these applications, both the sending and receiving of information matter, as they require extremely fast—and reliable—response times, sometimes even under a millisecond. These types of applications also necessitate a dense network coverage to make sure the response is always sufficiently quick.

In addition to speed and density of the network, device-to-device communication can enable superfast response when devices directly interact with each other. Also, edge computing brings analytics closer to the user's location, and new ways of multiplexing can ensure that mission-critical transmissions in the network are prioritized. Potential applications include mobility services, such as autonomous drones or vehicles and remotely operated services like remote surgeries.

Massive Machine Type Communications

Massive machine type communications (mMTC) is about direct interaction among large numbers of devices and computers without the intervention of people. It is defined as fully automatic data generation, transfer, analysis, and actuation among intelligent machines. Communication amongst machines is expected to generate the majority of traffic within 5G communication networks. Each device may generate small amounts of data at some intervals (e.g., temperature reading every hour), or collect and transfer a constant, real-time stream of data about the performance of an industrial machine. This use case probably requires less new capabilities from the communication network and more effort and innovation on the part of the industrial users to instrument and digitize their activities. Potential applications include sensor-enhanced production or operations, smart spaces (cities, homes, cruise ships), smart electricity grid, and remote control of equipment in remote or hazardous places.

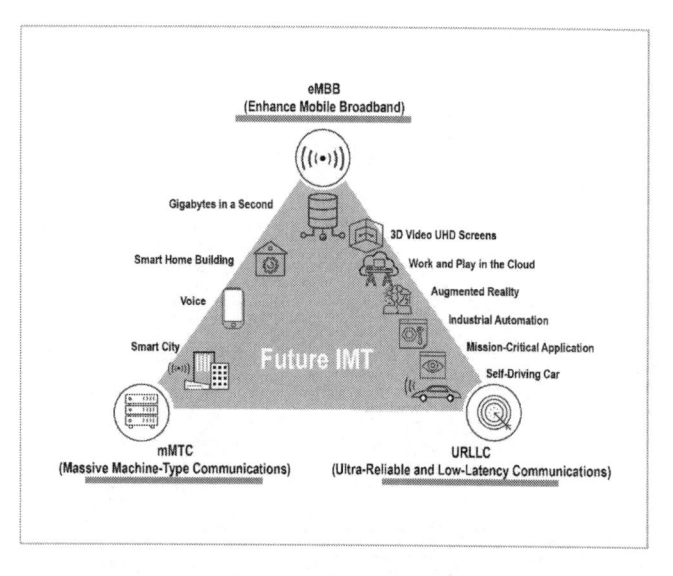

Figure 6.5. Future information management technology composition.

5G applications

Transformative 5G applications will exploit as many of the following 5G benefits as possible:

- *Low to zero latency*. When compared to 4G, 5G cuts the latency significantly (one millisecond). This leads to less network interference and delays. Lower latency rates are essential to the deployment of AR/VR based mobile apps.
- *Increased connection density*. 5G supports the connectivity of up to one million devices in the same physical location. This is a significant improvement over 4G and enables the broad use of IoT (Internet of things).
- *Improved precision*. With its higher radio frequencies and shorter wavelengths, 5G provides more precise geolocation capabilities.
- *Enhanced battery life*. Increased speed and reduced latency reduced device battery consumption. This extends the battery life of 5G devices by up to ten times.
- *Faster file transfers*. Through more efficient use of the electromagnetic spectrum, 5G will enhance file transfer speeds and virtually eliminate data transfer lag.
- *Next-level, media-rich user experiences*. 5G devices have vastly improved user interfaces. These improvements encourage more robust user interfaces.

- *Greater capacities and more Features.* Faster speed and ultralow latency give developers the opportunity to increase the number and capability of device features. User experience will be greatly enhanced, increasing application revenue and adoption.
- *3D models.* 3D gaming and immersive augmented reality will get a boost from 5G. Mobile apps coupled with 3D printers will enable additive manufacturing on a much broader scale than it is done today.
- *New generation of personalized chatbots.* Synergies between 5G and chatbots will significantly improve real-time interaction and feedback. It could also make the "tactile Internet" a reality with the deployment of haptic controls.
- *Less reliance on hardware.* 5G will reduce mobile application reliance on hardware because processing will be done in cloud datacenters.

Challenges associated with successfully developing, deploying, and operating new solutions include the following:

- *5G-based business model.* 5G network speed will unleash new 5G-based business models that will be designed to take full advantage of 5G networking capabilities.
- *Multiple app versions.* With 5G, mobile app developers will need to create multiple applica-

tion versions in order to maintain backward compatibility with 2G, 3G, and 4G networks.
- *Ambient computing.* 5G will enable ambient computing through which user digital experiences will include the ability to involve the surrounding environment and conditions with simultaneous interaction with machines, devices, sensors, and other advanced technologies.

Business Model Innovations

5G network features and capabilities open up a whole new world of new use cases that were impossible using 4G. Some of the most interesting prospective business models are covered below by industry vertical. The radar charts in the following sections highlight the 5G features that are central to delivering innovative solutions in the referenced industries.

Consumer Services

Augmented reality (AR) and machine learning (ML) technologies will drive the digitization of all industries. The wide deployment of interactive media applications will happen on 5G networks.

Transformative business models in consumer services include the following:

- Creating in-store 3D printers to create and deliver custom products to customers or deliver to customer-owned 3D printers

- Using VR and AR to virtually remodel and redecorate a customer home with in-store products or train employees in near real time
- Storefront able to sense customers and virtually cloth them in the latest fashions that match the customer's current and predicted preferences
- Robots empowering retailers to improve operational efficiencies and customer experiences by identifying out-of-stock, mispriced, or misplaced inventory as well as store hazards
- Retailers leveraging the physical store as a customer data source for delivering personalized experiences across channels and platforms, and 5G eventually delivering them in near real time
- Mobile apps loading remote virtual tellers, financial advisors, and loan officers, and using high-definition charts with personalized offers
- Bank customers able to see visually compelling stories about daily spend habits, compound interest, or the impact of impulse spending
- Stock traders able to visualize investments, trades, and budgets in 3D
- Marketing teams visualizing customer journeys in physical space using applications that map and simulate buy flows

- Compliance training programs for new hires transformed into immersive, media-rich sessions with animations overlaid onto real-world interactions
- The accelerated payment authentication and proactive fraud detection significantly improved with 5G. Mobile payments more secure through the use of cross-referenced merchant ID, transaction amount, geolocation, and biometrics
- Financial institutions able to extend their reach by providing a branch experience at temporary locations
- Remote risk assessment of commercial fleets enabled using ultra-HD (UHD) video
- 5G enabling beyond line-of-sight remote claims work and predictive risk analytics. Investigators overlaying government flood data on a mobile device to visualize potential damage
- Remote field appraisers assessing damage to a vehicle or home from a customer mobile device, pausing, capturing, and annotating UHD video for analysis and updates to HQ in near real time. Customers receiving on-the-spot settlements more quickly and accurately

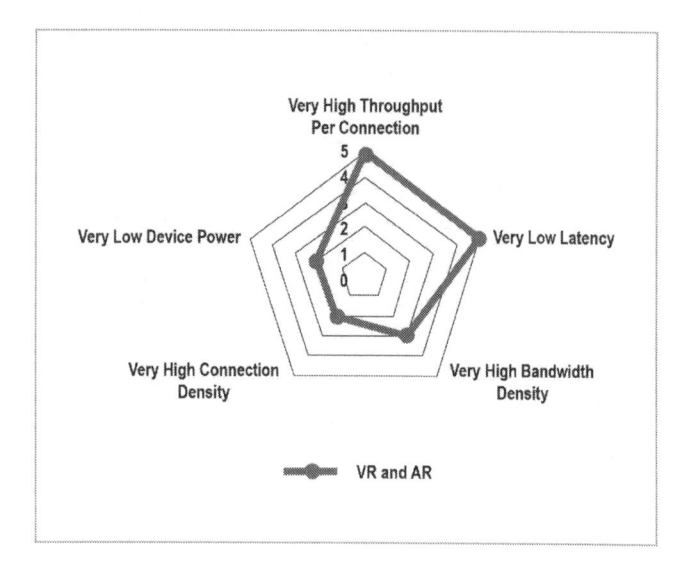

Figure 6.6. Network requirements of customer service applications[18].

Industrial IoT

Industrial automation and robotics are a very fertile area for innovation using 5G networking technology. Referred to as Industry 4.0, the three key components are these:

- Interoperability and connectivity within and beyond smart factories (an industrial Internet of things)
- Virtualization using deep sensor networks to monitor physical factories

---

[18] Source: https://www.huawei.com/minisite/5g/img/GSA_the_Road_to_5G.pdf

- Real-time capabilities for process control

These elements have important implications for the communication networks. Wireless sensor modules need secure, ultrareliable communication but also must have low power requirements. The larger number of connected devices may require low-latency links digital models, design engineering processes, and the physical plant.

The financial impact of IoT on factories in areas like as predictive maintenance and operations management could be between USD1.2 and USD3.7 trillion in 2025.

Transformative business models in industrial IoT include the following:

- Data transport over high-speed local area networks using small-cell technology to support network edge computing (NEC) and multi-access edge computing (MEC) solutions.
- Smarter integration and analysis of sensor-generated data using highly reliable—and highly secure—compute services.
- Increased factory optimization through AR and VR—that can deliver overlays that help guide workers through production steps for intricate assembly processes.
- The precise manufacturing tolerances using noncontact metrology. Using 5G, the data can reside at the edge of a network and simultaneously meet the demand for consistent

measurements and near-real-time big data analytics.

- Autonomous-guided vehicles (AGVs) that can process instructions and make intelligent decisions for safe navigation.
- Digital twin technology can give you a view into the most complex machines. In factory floor and operations planning, digital twins help engineers design layout and flow before making costly installations and changes.

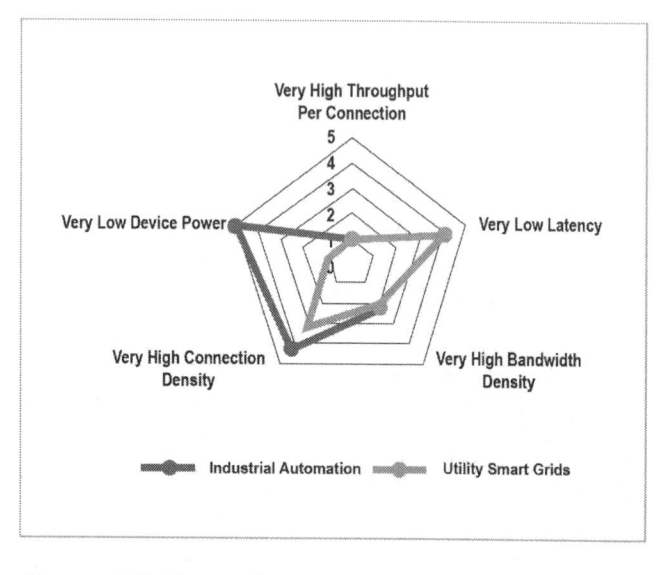

Figure 6.7. Network requirements across industrial and utility applications[19].

---

[19] Source: https://www.huawei.com/minisite/5g/img/GSA_the_Road_to_5G.pdf

Connected Transportation

Today, there are multiple vehicle-to-vehicle (V2V), vehicle-to-infrastructure (V2I), and intelligent transport systems (ITS) applications being rapidly developed. These near-term solutions require very low network latency. Driverless cars will also need real-time safety systems that can exchange data with other vehicles and the surrounding transportation infrastructure.

Collision avoidance systems will need 5ms latency and 99.999 percent reliability. Network economies of scale when solutions are built on public cellular networks will make these new autonomous transportation systems commercially feasible.

The connected vehicle market is expected to be very large. It will require a huge number of connected endpoints. Logistics, vehicle telematics, and automotive insurance industries will rely on these systems to improve their revenue and margins.

In addition to connected vehicles, there is a significant market for wireless 5G connectivity in ports, airports, railways, and shipping. These applications include large-file and real-time data exchange, and real-time information and entertainment systems.

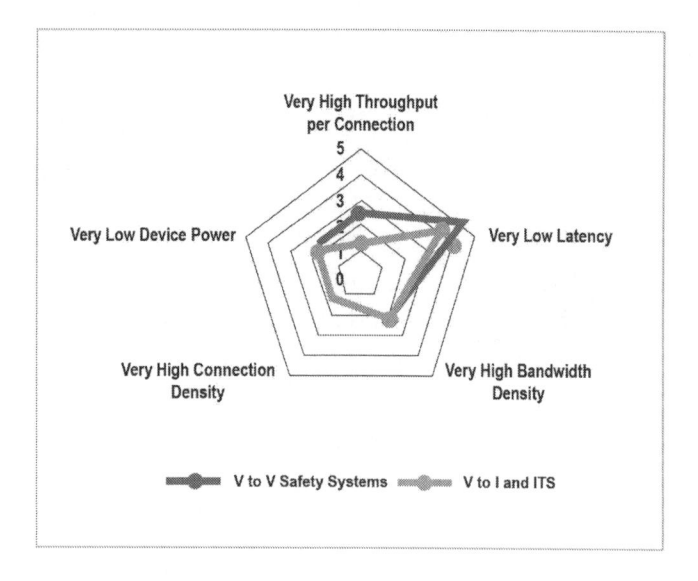

Figure 6.8. Network requirements for transportation applications[20].

Healthcare

Mobile health applications were given a significant boost during the COVID pandemic. Telemedicine concepts that had been developed over many years were suddenly and immediately put into active use. Personal health records, fitness data, and smartphone-based applications for the monitoring and treatment of long-term medical conditions have been widely deployed. The mobile network is a crucial tool for simulating and mapping virus outbreaks.

---

[20] Source: https://www.huawei.com/minisite/5g/img/GSA_the_Road_to_5G.pdf

5G will bring faster, more reliable, lower-latency networks to bear in the expansion of this current usage.

The development of remote, robotic surgery using VR, AR, and haptic technologies will also grow. Applications will include the following:

- *Telemedicine.* The use of language translators with video conference tools between a patient and doctor at the network edge with low latency.
- *Remote patient monitoring.* 5G will enable better connections on mobile devices, which will help enable improved live and real-time remote care. Wearables are actually predicted to decrease hospital costs by 16 percent over the next five years.
- *Augmented and virtual reality.* 5G-powered AR/VR can help doctors, nurses, interns, and staff better visualize procedures in a learning-by-doing setting.
- *Data analysis.* With 5G, data will be distributed to multiple points of care, enabling innovative early remote diagnoses, remote surgeries, smart hospitalization logistics, and improved patient engagement.
- *Decentralizing the healthcare model.* 5G will support the decentralized healthcare ecosystem by making operations more reliable and accessible.
- *Large file transfers.* 5G has the capacity to enable faster transfers of huge medical images. Faster image transfer will make patient care more efficient, and hospitals will be able to see more patients in less time.

The requirements of wireless health applications are summarized in figure 6.5.

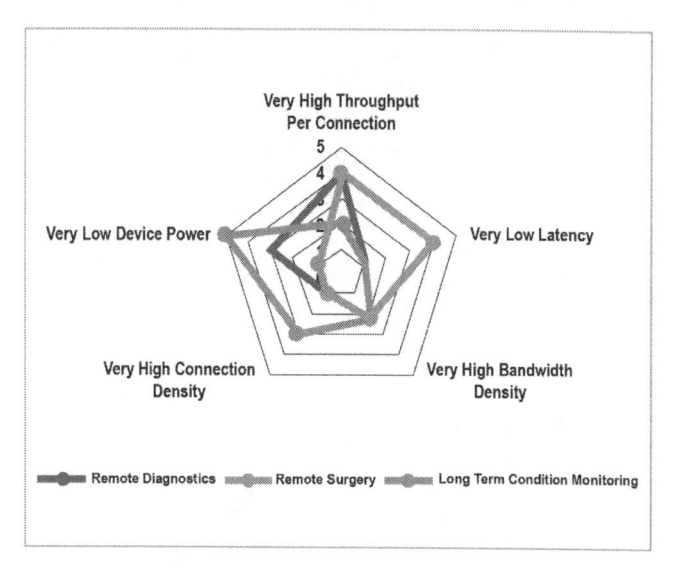

Figure 6.9. Network requirements for healthcare applications[21].

Smart Cities

Smart city applications use enhanced network connectivity and IT infrastructure to deliver operational processes that encompass greater levels of intelligence and automation. Many compelling examples aggregate data from multiple domains (transportation, public administration, emergency services,

---

[21] Source: https://www.huawei.com/minisite/5g/img/GSA_the_Road_to_5G.pdf

weather sensing, etc.) to deliver better planning and real-time automated responses to dynamic situations.

Smart cities generally use deep-sensor networks. Wireless network technologies lend themselves well to providing the needed connectivity. The most ambitious smart city ideas need ubiquitous connectivity through a network capable of supporting multiple dataflows with varying performance requirements as some smart city applications are "mission critical."

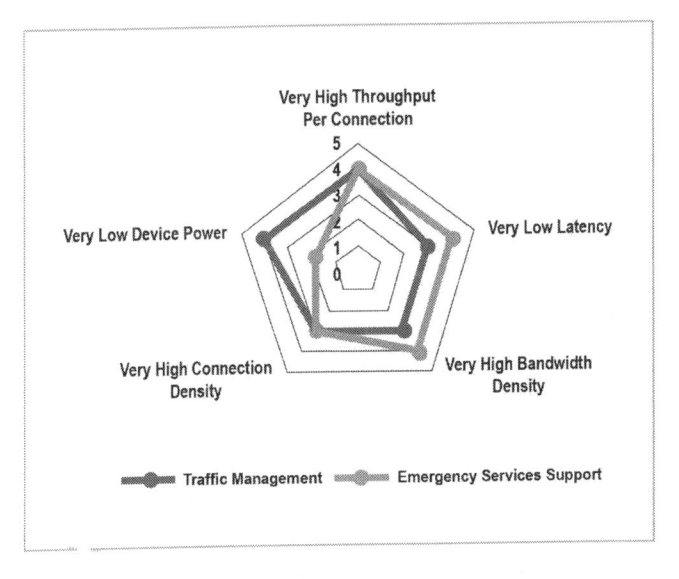

Figure 6.10. Network requirements for smart city applications[22].

---

[22] Source: https://www.huawei.com/minisite/5g/img/GSA_the_Road_to_5G.pdf

5G Business Ecosystem

New 5G solutions require business ecosystems. All required products and services must be fully available at the time of launch or you risk delay or poor customer experience.

Ecosystem participants include the following:

- *Nonmarket participants.* These parties are a part of the 5G ecosystem and influence it indirectly by creating regulations and laws that impact the 5G ecosystem.
- *Standard development organizations.* These are industry-specific, national, or international associations that facilitate and coordinate the development of communication standards.
- *Governments institutions.* These govern nations and create and enforce laws and regulations that can impact both markets and technological infrastructures.
- *Market participants.* These participants are central to the 5G ecosystem. The primary functions of these commercial organizations surround 5G, whether it's in developing technology for 5G or facilitating 5G networks.
- *Telecommunication operators.* These parties often finance the investment into telecom networks and provide ongoing services and network management.
- *Network equipment providers.* These high-technology companies invest in technology development, standard development, and patenting,

and then build telecommunication networks on behalf of telecommunication operators.
- *Terminal manufacturers.* These manufacturers create devices that communicate within the network. These devices can range from on-board computers on vehicles to embedded mobile chips and sensors.

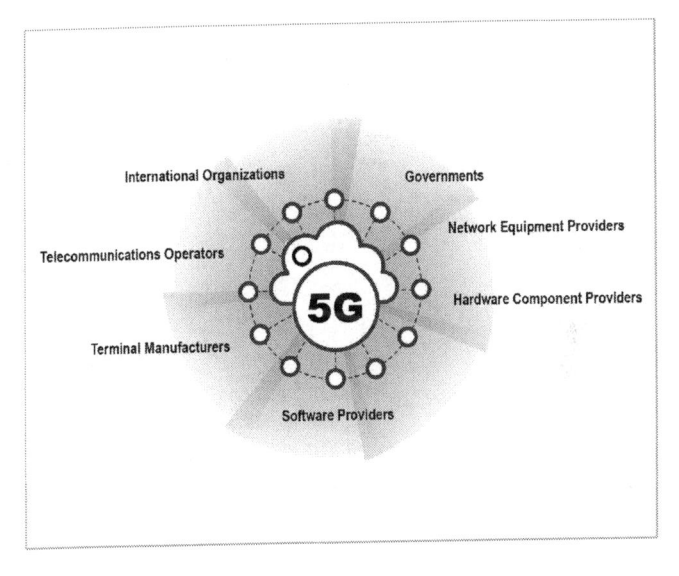

Figure 6.11. 5G business ecosystem.

The key ecosystem challenge is balancing cooperation amongst the members with control of the solution components by the members. Often described as "coopetition," this friction is one of the most prevalent causes of business model failure. A frequent root cause of this is intellectual capital disputes and patent licensing disagreements.

Cooperative Standardization Model

Information and communications technology (ICT) markets rely on standardization and interoperability. This is required to achieve economies of scale, high efficiency, and network effects. The low-cost, high-quality transmission of data relies on the infrastructure supplied by the organizational and contractual arrangements that drive market standardization. Standardization will also be a requirement for 5G deployments. When the 5G network and associated "Internet of things" are deployed, the emerging environment is expected to follow the successful implementation of the 2G/GSM, 3G/UMTS, and 4G/LTE standards. The 5G network will comprise data exchange not only between individuals but also between devices. This shift will be both a technological and organizational challenge.

Referred to as the cooperative standardization model, this market institutional structure has been targeted by antitrust agencies in the United States and other commercially significant jurisdictions. This regulatory campaign has culminated in the landmark antitrust litigation brought by the Federal Trade Commission (the FTC) against Qualcomm, a semiconductor firm widely recognized as the leading innovator in the smartphone industry. Prior to and concurrently with the FTC suit, competition regulators in the US, China, the European Union, South

Korea, and Taiwan have also sought to limit a firms' ability to enforce patents and licenses in the smartphone market.

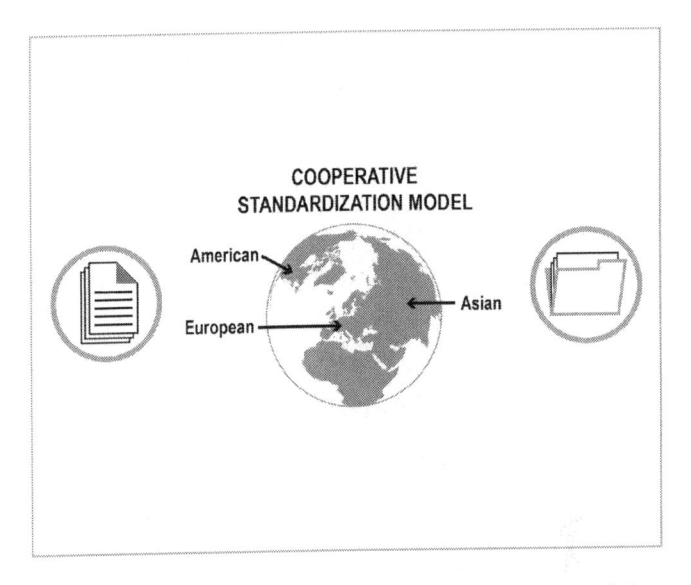

Figure 6.12. Cooperative standardization model.

No matter the final legal decision in each country, this activity will regulate how companies generate market power by blocking competitors' use of patents or collecting royalties from competitors. Open standard development organizations and patent pools help regulate the terms under which standard, essential patents must be licensed. This makes participation in standards bodies a key component of any transformational business strategy.

The Innovation Opportunity Framework

The Cornell University innovation opportunity framework[23] can help companies assess potential innovation opportunities in the 5G context. This tool structures an evaluation of how the new 5G capabilities can generate improvements to current products and services, or alternatively, it can help in the development of brand-new products and services. The focus of the framework is on enhancing the user experience with 5G capabilities.

The first step is recording a general description of the innovation opportunity being evaluated. The description should then be evaluated on how it could be realized across each of the six dimensions of the innovation opportunity framework:

- *Client.* Who is the user, an individual consumer, or an industrial company? This determines whether the innovation requires a consumer telecom or a private business 5G network and how the service is launched in the marketplace. Who is the client who would pay for the service, if different from the user?
- *State-of-the-art.* How is the activity or service currently provided, and what are its performance issues or bottlenecks? In other words, what is the "pain point"?

---

[23] https://www.ecornell.com/courses/technology/innovating-with-5g/

- *Content.* What type of content will be transferred in the network: Is the network primarily about instrumentation—that is, data collection and real-time transfer using lots of devices (emphasis on the uplink), about network speed, and data volume—that is, downloading and streaming (emphasis on the downlink), or about latency and reliability—that is, involving rapid interaction among the users/participants (emphasis on both uplink and downlink)?
- *Mobility.* Do the services entail mobility or just depend on the quality of service in a given location? In other words, what density, coverage, and geographic area are needed?
- *Timing.* How important is latency? Is the content or transfer time-sensitive, or can it tolerate delays?
- *Investment.* Who will finance, build, and maintain the 5G network, a telecom operator or an industrial firm? What is the anticipated capital expenditure?

Based on an assessment of an innovation opportunity across these six dimensions, the business must then evaluate the revenue potential in this innovation opportunity to create unique, added value for users when building on 5G capabilities. Output from this analysis could serve as valuable input to a subsequent solution business case.

Technical Metrics

Once a baseline business case is established, the solution designer should then identify the technical metrics that can be used to evaluate progress toward attaining the desired business goal. ROI models should document how hybrid infrastructure solutions would benefit both business and IT consumers and providers. This exercise should also involve examining the key technology features and business operating model changes.

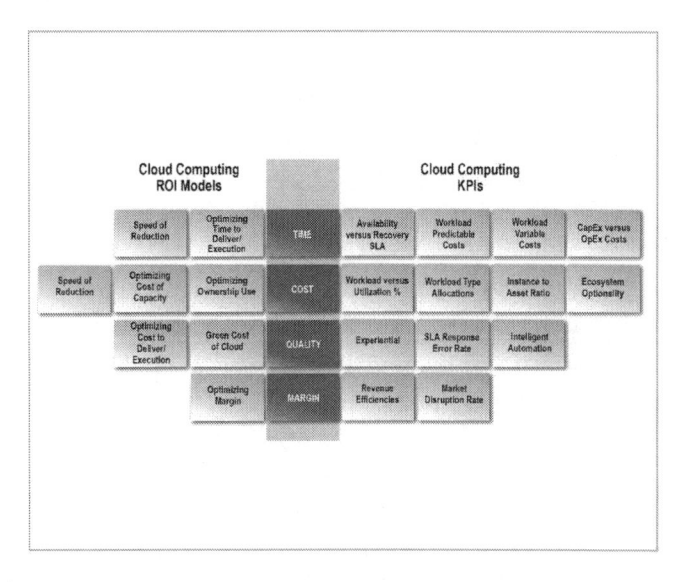

Figure 6.13. Cloud computing ROI models and KPIs.

ROI models to support hybrid solution assessments and business cases should review the following two aspects:

- Key performance indicator ratios that target cloud computing adoption, comparing specific metrics of traditional IT with cloud computing solutions. These have been classified as cost, time, quality, and profitability indicators relating to cloud computing characteristics.
- Key return on investment savings models that demonstrate cost, time, quality, compliance, revenue, and profitability improvement by comparing traditional IT with cloud computing solutions.
- The overview of cloud computing ROI models considers both indicators and ROI viewpoints.

Cloud ROI Cost Indicator Ratios

The figures below show the cost indicator ratios, and outline explanations are given following.

Cloud Computing ROI Models

Availability versus Recovery SLA	Indicator of availability performance compared to current service levels
Workload – Predictable Costs	Indicator of Capex costs on-premise ownership versus Cloud
Workload – Variable Costs	Indicator of Opex cost for on-premise ownership versus Cloud
CapEx versus OpEx Costs	Indicator of on-premise physical asset TCO versus Cloud TCO

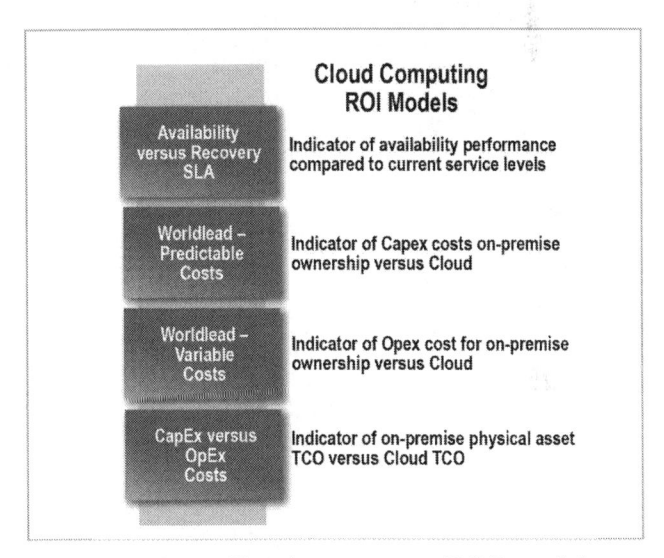

Figure 6.14. Cloud computing ROI models.

Click to Transform

- Availability versus recovery SLA—Indicator of availability performance compared to current service levels
- Workload—Predictable costs—Indicator of CAPEX cost on-premise ownership versus cloud
- Workload—Variable costs—Indicator of OPEX cost for on-premise ownership versus cloud; indicator of burst cost
- CAPEX versus OPEX costs—Indicator of on-premise physical asset TCO versus cloud TCO

Cost Indicator Ratios

Workload vs Utilization %	Indicator of cost=effective Cloud workload utilization
Workload Type Allocations	Workload size versus Memory/Processor distribution. Indicator of % IT asset workloads using Cloud
Instance to Assist Ratio	Indicator of % and cost of Rationalization/Consolidation of IT assets. Degree of complexity reduction (%)
Ecosystem Optionality	Indicator of number of commodity assets, APIs, Catalog items, self-service

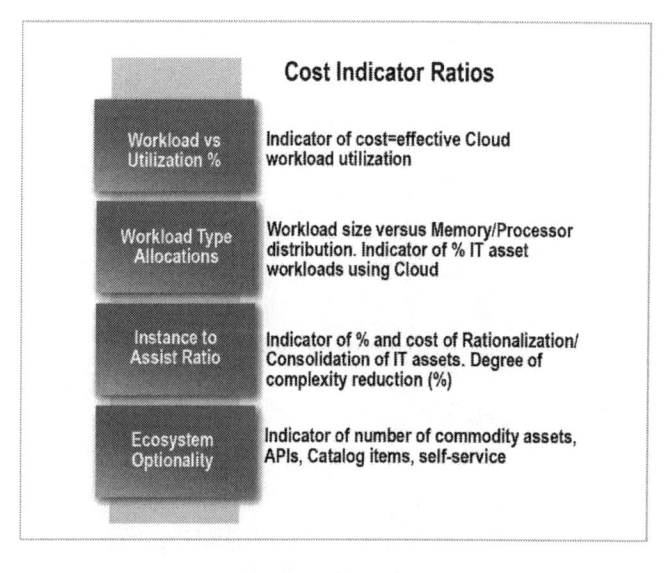

Figure 6.15. Cloud computing cost indicator ratios.

- Workload versus utilization percentage—Indicator of cost-effective cloud workload utilization

- Workload-type allocations—Workload size versus memory/processor distribution; indicator of percentage and IT asset workloads using cloud
- Instance to asset ratio—Indicator of percentage and cost of rationalization/consolidation of IT assets; degree of complexity reduction
- Ecosystem optionality—Indicator of number of commodity assets, APIs, catalog items, self-service

Cloud ROI Time Indicator Ratios

Figure 6.12 shows the time indicator ratios, and outline explanations are given below.

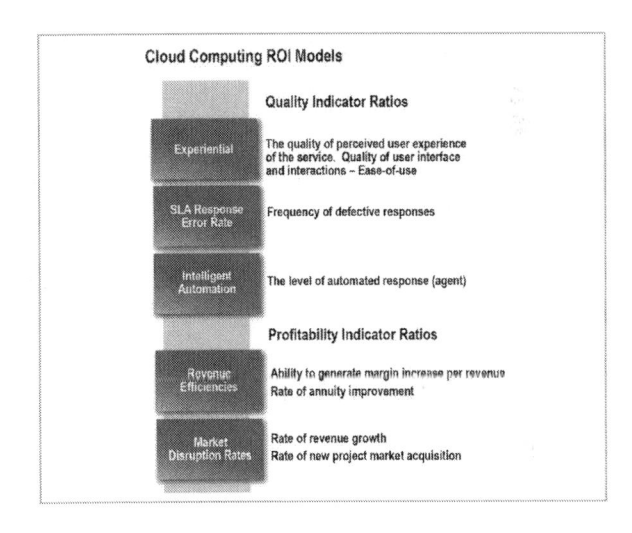

Figure 6.16. Cloud computing ROI models: quality indicator ratios and profitability indicator ratios.

- Experiential—The quality of perceived user experience of the service and the quality of user interface (UI) design and interaction and ease of use
- SLA response error rate—Frequency of defective responses
- Intelligent automation—The level of automation response (agent)

Cloud ROI Profitability Indicator Ratios

Figure 6.12 above shows the profitability indicator ratios, and outline explanations are given below.

- Revenue efficiencies—Ability to generate margin increase/budget efficiency per margin and Rate of annuity revenue improvement
- Market disruption rate—Rate of revenue growth and new product acquisition

Cloud ROI Savings Models

Figure 6.13 shows the savings models, and outline explanations are given below.

Cloud Computing ROI Savings Models

TIME	COST
Speed of Reduction	Speed of Reduction
Rate of change of TCO reduction by Cloud adoption	Rate of change of TCO reduction by Cloud adoption
Optimizing Time to Deliver/ Execution	Optimizing Cost of Capacity
Increase in provisioning speed / Speed of multi-sourcing	Aligning cost with usage, CapEx to OpEx utilizations pay-as-you-go saving from Cloud adoption
	Optimizing Ownership Use
	Portfolio TCO / License cost reduction from Cloud adoption / Open Source adoptions / SOAre-use adoption

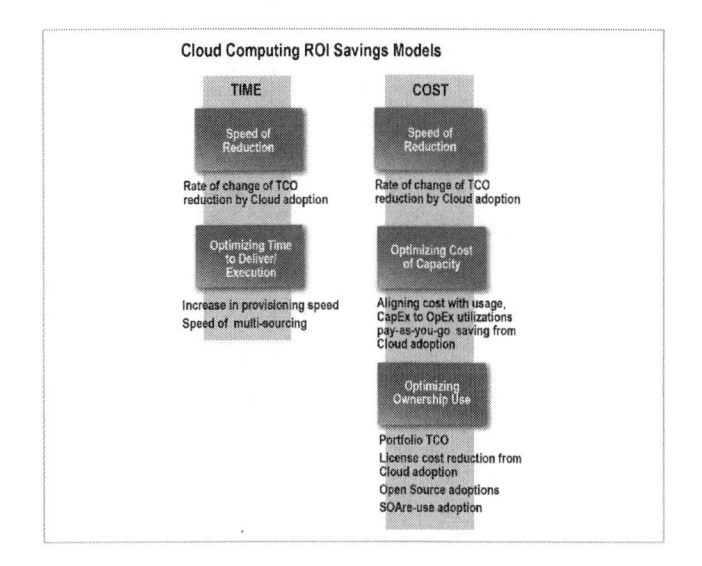

Figure 6.17. Cloud computing ROI savings models: time and cost.

- Speed of time reduction—Rate of change of TCO reduction by cloud adoption
- Optimizing time to deliver/execution—Increase in provisioning speed and speed of multi-sourcing
- Optimizing cost of capacity—Aligning cost with usage, CAPEX to OPEX utilization pay-as-you-go savings from cloud adoption
- Optimizing ownership use—Portfolio TCO, license cost reduction from cloud adoption. Open Source adoption. SOA reuse adoption

Cloud Computing ROI Savings Models

QUALITY

Green Cost of Cloud

Rate of change of TCO reduction by Cloud adoption

PROFITABILITY

Optimizing Margin

Rate of change of TCO reduction by Cloud adoption

Optimizing Cost to Deliver/ Execution

Increase in provisioning speed
Speed of multi-sourcing

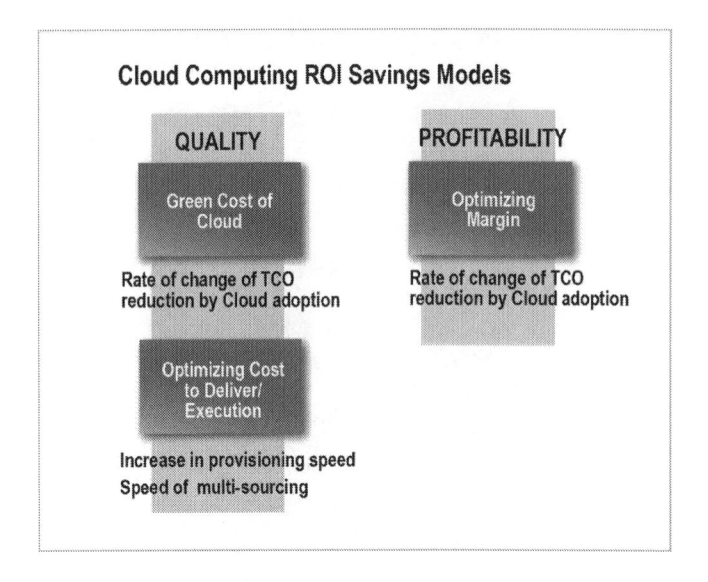

Figure 6.18. Cloud computing ROI savings models: quality and profitability.

- Green costs of cloud—Green sustainability.
- Optimizing time to deliver/execution—Reduced supply chain costs. Lower error rates
- Optimizing margin—Increase in revenue/profit margin from cloud adoption

Operational Metrics (the -ilities)

The -ilities are central to any discussion of engineering systems, and they represent the desired properties of systems, such as flexibility or maintainability. These usually present themselves after a system is initially deployed. These properties generally do not represent the primary functional requirements of a system but normally concern wider system impacts to time and user stakeholders than are embedded in those primary functional requirements. The -ilities do not include factors that are always present, including size and weight. The top four -ilities are quality, reliability, safety, and flexibility.

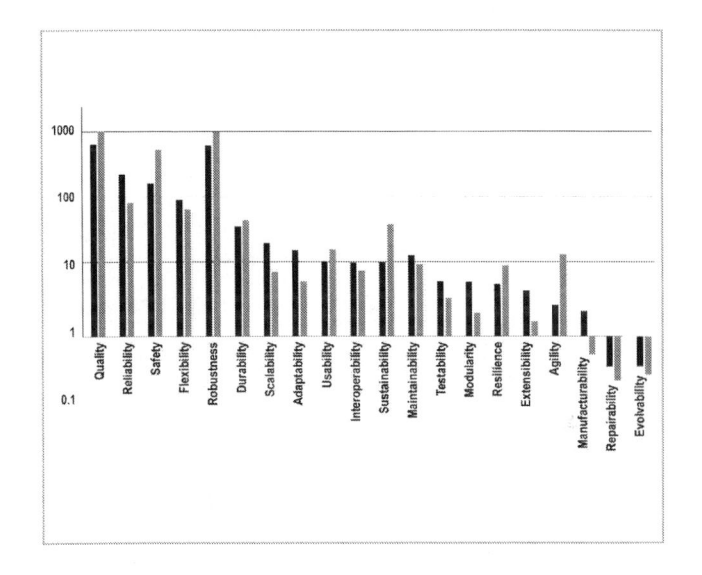

Figure 6.19. The ranking of the -ilities.

Additional -ilities have emerged as a result of growing complexity as well as the overall scale of deployments led to more and more important side effects. The increasing rate of change in systems and society has also spurred this expansion. Early system buyers did not see the value of paying for capabilities that did not directly support primary system functionality, but over time, it became untenable to run systems without paying attention to these other characteristics. Now, there is a realization that much of the value that systems generate depends on the degree to which they address certain lifecycle properties or -ilities. Attention should be focused on designing all possible -illities into any proposed transformational solution. As depicted in Figure 6.16, these system characteristics interact with each other in a reinforcing mesh, improving overall system value and user satisfaction. The following, however, should take priority.

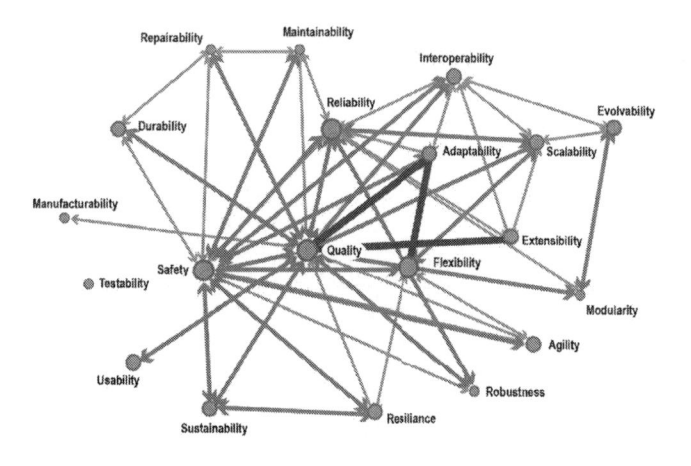

Figure 6.20. The interactive enmeshing of -ilities.

Quality

Quality is often described as a transcendent (some abstract philosophical, perceptual, moral, or religious entity), product based (fit for use, performance, safety, and dependability), manufacturing based (conforming to engineering and design specifications), user based (able to satisfy human needs), or value based (difference between conforming to specifications and monetary cost). In this context, quality means that the solution or service is well-made to achieve its function. Such quality is a direct result of system "tolerance." Technical metrics, described earlier, can be used to design in quality.

Safety

Safety is generally related to the avoidance of any harm to life. Understanding and evaluating the safety of any system requires a deep understanding of detrimental interactions among user behaviors, technical components, and the operating environment. As such, the specific use case will determine the relative importance of safety in any solution. For instance, safety matters a lot in transportation-related systems, but it doesn't come up nearly as often when we look at communication systems unless the absence of effective and reliable communication could result in harm.

Maintainability/Reliability

The fourth -ility is maintainability. Along with its counterpart reliability, both are related to quality, usability, and operability. There are two types of maintainability. The first, preventive, describes the care taken to ensure than an artifact or system doesn't break down. It might mean replacing some parts before damage from wear and tear gets too bad. The other type is corrective, which involves repairing things that break to restore the artifact or system to its fully functioning state.

Usability/Operability

When referring to complex systems, human factors emerged as the key contributor to usability or operability. Although sometimes considered synonymous, they are different. Usability closely corresponds to human factors and ergonomic issues, whereas operability more clearly denotes institutional concerns beyond the scope of single human beings.

Flexibility

Flexibility describes the relative ease with which a system can be changed to embrace change or interact with another system. This characteristic is also related to reconfigurability, the ability to change into different configurations that allow the system to perform multiple functions. Flexibility also functions as an umbrella term for a number of other related -ilities. Evolvability regards fundamental changes made to a system's very purpose. This tends to be manifested over the long term and involves deliberate initiatives. Modularity can be seen as an important prerequisite for flexibility.

Adaptability, by contrast, is a belief that changes in the system are driven by changing external environments. Striding both regimes of flexibility, operation, and redesign, an adaptable system is one that can be reconfigured in response to external stimuli. Both are related to the ability of a system to change rapidly, referred to as agility. Two other -ilities under the flexibility umbrella are scalability and extensibility. Scalability is the ability to grow the size of a system to support a greater number of something, be it how many users the system supports or how many daily transactions can be completed by the system. Scalability is about volume and involves both operations and redesign. Extensibility is about extending the way a system works so that it can fulfill multiple functions. A system can be flexible in some areas and inflexible in others.

Resilience

Resilience describes how fast a system can recover from a major disruption while regaining its original level of performance. That recovery may mean adjusting during the disruption or soon thereafter so that the system can sustain its required operations under expected or unexpected conditions. Where designing for flexibility involves more proactive planning for possibilities, designing for resilience is about creating a system that can bounce back from something unexpected.

Like flexibility, resilience is an umbrella term. Elements of adaptability speak to a system's resilience. Robustness, for example, is the ability of a system to work as intended when conditions change.

Interoperability

The last of our primary -ilities is interoperability. This characterizes systems that can function independently but can also work as part of a larger whole. Related to interoperability are compatibility and modularity. Compatibility tends to relate to consumer products and systems and describes how well components of the system can be connected and work together.

Modularity has two primary aspects. One is functional decomposition, and the other is encapsulation. Functional decomposition means that the subfunctions of the larger system can be decomposed and assigned into smaller units or "modules" of the larger system. Encapsulation describes the ease with which a system can be pulled apart and reassembled. Having both aspects provides the greatest level of modularity, which is a powerful enabler of interoperability and other -ilities.

Summary

Operational metrics were not always on my mind as I approached new projects. When I reflect on my first company, I see a man hyper-focused on technology. If I had taken a step back to understand basic operations and business models, maybe Zocom would have benefitted a larger audience. Instead, I only look back and see a failed partnership with Xybernaut and an embarrassing product. Many people explain the failure of The Mobile Assistant to the cliché that the product was "ahead of its time." The Mobile Assistant was not ahead of its time. Although it was a wearable computer, the approach to marketing and business operations was far from innovative. That's why it failed to transform.

Yes, wearable computers did exist in 1996. Instead of the "wearable computers" we think of today (Apple Watch, Google Glass, etc.), the Mobile

Assistant resembled an actual PC worn as clothing. Users donned a utility belt and headpiece containing a hard drive, mini monitor, batteries, and even a mouse. It was as clunky as it sounds. No amount of innovative working technology within the Mobile Assistant could overshadow the fact that users were weighed down by their gadgets. Users also looked like total dorks.

At the time, I didn't care that users looked goofy. My mind was so focused on the technology that I overlooked any practical or operational roadblocks in my way to transformation. I had accomplished the first step in achieving transformative innovation by creating something innovative. I believed that the market would do the rest for me—and that's where I failed.

You alone cannot determine if a product or service is a transformational innovation. Objectively, your product can be innovative, like The Mobile Assistant. To be transformational, your product must serve a need in a large marketplace that will accept this grand transformation.

Let's think back to Airbnb. Over 40 million Americans checked into an Airbnb in 2019. The process for listing and booking private property was certainly innovative, but more importantly, it appealed to the needs of a market that wanted a unique alternative to hotels and motels. This market was large enough to transform and disrupt the industry. Uber? The same idea applies. The market of people who wanted a more convenient alternative to taxis (and the market of people who wanted to make some extra cash) was large enough to transform the rideshare industry.

The number of people in the market for a dorky, hard-to-manage wearable computer paled in comparison. Xybernaut sold less than 10,000 units of the Mobile Assistant between 1995 and 2005. (Compare that to the 31 million Apple Watches sold in 2019.)

"Clunkiness" wasn't the only roadblock standing in the Mobile Assistant's way toward becoming a transformational innovation. None of the -ilities were present in the Mobile Assistant or the way that it was produced. The entire system behind the product lacked the quality, reliability, and scalability required to transform. Even if the market was large enough to create demand for the Mobile Assistant, it's unlikely that the product would be a commercial success.

After the failure of the Mobile Assistant, I took a step back and saw a much larger picture. If I had seen this picture before, I could have predicted the failure of a wearable computer in the late '90s. Never again would I focus so intently on the technology and fail to consider operations, the consumer market, and big-picture results. This partnership marked a change in the way that I approached business and the future of cloud computing.

7

Transformation Frameworks

In the beginning, digital transformation was a journey through a tropical jungle with no paths to follow, no map or compass to use, or a solid destination to find. Today, however, research on the successes and failures of those that entered the jungle before you are available. Although the journey can still be difficult, the following transformation frameworks can enhance your chance of success.

Harvard Business Review

An *MIT Sloan Management Review* article described three pillars and nine building blocks for digital transformation. These are meant as steppingstones you can follow towards digitization if your business.

Transform Customer Experience

According to Harvard, organizations should first improve their customer understanding through the use of analytics. In this model, data on customer behavior defines market segmentation and the appropriate targets for company products and services. Commodity cloud services can now make that task much easier and less expensive through big data analytics services. Historical data can now be quickly processed for important business defining nuggets.

This analysis often provides critical insights into the behavior and preferences of different market segments and geographies. Companies can also analyze social media signals and online discussions to gather real-time insights. This data source can complement historical data to provide an overall view of the market.

Enhancing customer experience through mobile device interactions can also lead to top-line revenue growth. This is often accomplished through optimizing the sales process or physical buying experience. Cloud computing is normally the enabler for these types of transformations. Cloud can also be used to empower sales representatives, improve vendor partner interactions, and expand brand image. Digitizing customer touch points and delivering an omnichannel experience to customers are also important options to consider.

Transform Operational Processes

Automation is often key to business process digitization. Often a focus on operations can lead to business process optimization and entirely new revenue streams. Three common routes for this are the following:

- Robotic process automation (RPA) which uses artificial intelligence to automate routine activities
- Remote worker enablement that uses workspace virtualization and advanced visualization technologies
- Performance management decision making based on real-time data that delivers deeper insights into customers, products, and operational regions

Transforming Business Models

Remodeling of business refers to the change in the method of interaction between departments and customers. Many new digital services are being created by augmenting traditional physical products with digital offerings. In retail, this is often done by using digital media to drive new business. Other businesses like food and beverages, fashion, and industrial goods manufacturers are incorporating digital mediums into their current business process. Another method of establishing new digital businesses is by redefining the digital boundaries of current operations. Digital globalization and the use of global shared services to promote efficiency, from multinational to global presence, can also deliver real value.

Cognizant

The Cognizant framework digitizes process components with a goal of integrating them into a successful end-to-end process.

Digitizing the Customer Experience

Customers are the most important stakeholder of any organization. To serve customers in the best way possible, it's essential to understand them. Until recently, this was only possible through the use of a company's internal systems.

Using CRM, companies could identify which products a particular customer has historically purchased. Now, a better and broader analysis can be performed using the Internet. By observing and recording the digital footprint of consumers, information can be accumulated on customers, processes, organizations, and devices. Based on information derived from social media, organizations can link consumer preferences to buying habits. Every action that consumers take, in both the physical and virtual worlds, contributes to the digital footprint value.

Customer insights acquired using digital technology can also be used for digital marketing. The first thing that consumers and business customers do today when researching a purchase is to check the Internet, visit websites, and consult their personal networks and communities for advice and rankings.

Competitive organizations provide up-to-date product information online and engage with online communities to provide advice on their products. Since customers and businesses are increasingly active and identifiable online, organizations can use digital marketing tools to personalize their product and service promotions, with the goal of increasing customer loyalty. With new channels of interaction, such as mobile and social media, customers now expect all their engagements with the company to be consistent across all available channels.

For instance, they expect to place an order when and where it's most convenient for them and then to receive their products through the channel of their choosing. If organizations do not enable such an omni-channel approach, they risk losing customers and increasing customer dissatisfaction. Omni-channel communication and service is, therefore, key. It is no longer a question of whether your organization should act on this but when, as consistent and efficient interactions across channels are now a crucial requirement for competing in the digital world.

Digitize Products and Services

Organizations today increasingly realize they can no longer focus on just selling products; they need to sell an experience. An increasing number of products today both consume and generate data, and many are interconnected through the web. Because of this increased intelligence, their usage can be monitored, additional services can be proactively offered, or maintenance can be provided when a problem is detected.

A good example is a smart toothbrush with sensors scanning your teeth for any problems. The physical tool itself is a commodity, but a user's brushing habits, dental hygiene history, and health needs create a Code Halo of information of premium value. This information can be sent to the dentist, who can provide feedback and advice or schedule an appointment.

The Code Halo can also be used by the device manufacturer to make product improvements. As it becomes easier and more affordable to make products Code Halo-capable, the question will soon become not what to wrap with a Code Halo but what not to wrap with a Code Halo. To create such experiences, customers should be taken on a journey through the process. To accomplish this, organizations need to think about how they can stay in touch with customers throughout the entirety of their journey.

Digitizing Operations

Advanced digital technology, powered by the SMAC Stack and aided by sensors, can improve business processes in several ways. For example, big data analytics can help inbound logistics run more smoothly by tracking product movements; the cloud can be used to create uniform business processing platforms, and mobile platforms can enable employees to perform their work anytime, anywhere, on any device.

Standard cloud platforms offer features and functionality updates more quickly and can lower testing costs. When using standardized platforms, either within a large organization (as a propriety platform) or as an open-market standard, it is also much easier to globally source processes, which can lead to substantial cost reductions.

Based on our own experience, aggressive sourcing and off-shoring can cut up to 50 percent of operational costs. More organizations are choosing to do this across a larger variety of business processes and IT services. By automating, standardizing, and globally sourcing processes, organizations can become more agile, more responsive to changes in demand, and better able to increase and sustain profitability. Such agility is essential as competitiveness is increasingly dependent on responding and anticipating to fast-changing market developments through human intervention; artificial intelligence and automated machines are not yet fully able to respond. Therefore, organizations must adopt an agile way of working.

Digitizing the Organization

With value chains increasingly integrated among businesses, organizations can become part of a larger ecosystem, enabling them to offer end-to-end services to their customers.

Insurance companies can, for example, offer a car replacement when a customer's automobile breaks down, improving the customer experience through this added service. Digital solutions can support value chain players to work more closely together. Organizations can either shape and orchestrate an ecosystem themselves and provide a significant number of products and services, or focus on a niche service that adds value to the customer experience and becomes part of an already existing ecosystem.

Organizations that shape and orchestrate an ecosystem and introduce their standards into the industry value chain, like Airbus and Walmart, tend to become dominant players in global markets. To work effectively within an integrated ecosystem, employees need to work together in a new way, breaking down silos and collaborating across different departments.

Employees need to learn from each other in order to respond more quickly and consistently to changes in the market and within their own organization. Geographically dispersed employees need collaboration tools to share documents, ideas, contacts, experiences, and knowledge so that they avoid "reinventing the wheel" and provide the business with enhanced value.

Corporate cultures also need to move toward a digital mindset; innovation should be rewarded, and additional digital expertise can be brought in to help employees embrace the digital world and acquire the necessary skills and knowledge. Digital collaboration today stretches beyond the borders of the organization, with communities cocreating products or services, and customers providing opinions and suggestions for product improvements through online forums. In this way, customers can influence product development, benefiting both themselves and the business.

Altimeter

Altimeter framework features a six-stage maturity model for digital business transformation. This approach includes the following:

Business as Usual

Organizations operate with a familiar legacy perspective of customers, processes, metrics, business models, and technology, believing that it remains the solution to digital relevance.

Present and Active

Pockets of experimentation are driving digital literacy and creativity, albeit disparately, throughout the organization while aiming to improve and amplify specific touchpoints and processes.

Formalized

Experimentation becomes intentional while executing at more promising and capable levels. Initiatives become bolder, and as a result, change agents seek executive support for new resources and technology.

Strategic

Individual groups recognize the strength in collaboration as their research, work, and shared insights contribute to new strategic road maps that plan for digital transformation ownership, efforts, and investments.

Converged

A dedicated digital transformation team forms to guide strategy and operations based on business and customer-centric goals. The new infrastructure of the organization takes shape as roles, expertise, models, processes, and systems to support transformation are solidified.

Innovative and Adaptive

Digital transformation becomes a way of business as executives and strategists recognize that change is constant. A new ecosystem is established to identify and act upon technology and market trends in pilot and, eventually, at scale.

The Six Stages of Digital Transformation

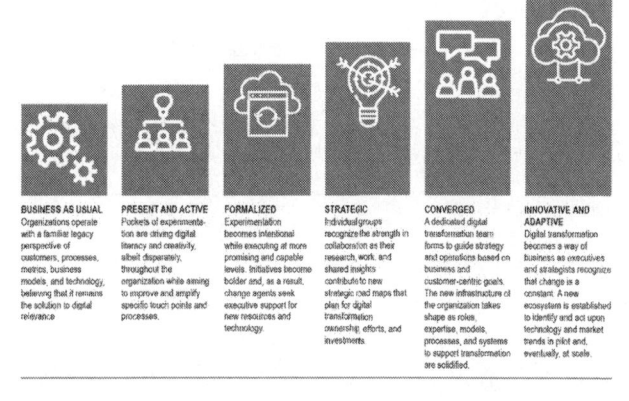

Figure 7.1. The six stages of digital transformation.

Collectively, these phases serve as a digital maturity blueprint to guide purposeful and advantageous digital transformation. Altimeter research of digital transformation is centered on the digital customer experience (DCX) and thus reflects one of many paths toward change. They found that DCX was an important catalyst in driving the evolution of business, in addition to technology and other market factors.

- *Governance and leadership.* An infrastructure that is driven by leadership philosophies that determine the fate of the business evolution.
- *People and operations.* Who is involved? Who is involved in the digital transformation (DT)? What are the roles they play, the responsibilities and accountabilities they carry, and how does a company enact change and manage transformation, including its roles, processes, systems, and supporting models?
- *Customer experience.* The processes and strategies aimed at improving touch points along the entire customer journey.
- *Data and analytics.* How does a company track data, measure initiatives, extract insights, and introduce them to the organization?
- *Technology integration.* Implementing technology that unites groups, functions, and processes to support a holistic CX.
- *Digital literacy.* Ways in which expertise is introduced to the organization.

According to the research, understanding the six stages of digital transformation maturity will lead to the following business benefits:

- Customization the maturity model to inform specific road map development
- Peer company benchmarking
- Executive alignment and buy-in
- Bolster sense of urgency
- Future marketing trend insights
- Prioritize digital transformation initiatives
- Set a new vision, course, and platform for leadership
- Develop new models, processes, and a purpose for technology and the future of work

Ionology

Establish a Digital Strategy

Use a framework to manage the process of creating a digital business strategy. We recommend the following seven principles as outlined in Niall McKeown and Mark Durkin's *The Seven Principles of Digital Business Strategy*. There are usually only a limited number of clearly defined plays a business can make. Selecting the right play is critical if the business is to maximize its resources, focus on what makes it effective, and move its market position.

1. *Know yourself.* A clear diagnosis of the situation, an easily understood strategic ambition, and a well-articulated value proposition.
2. *Customer.* An empirical value for customer volume, the tasks they wish to complete and their motivations.
3. *Marketplace.* Who is the competition, and what is their momentum?
4. *Resources.* Time, talent, and cash can be dedicated to fulfilling the strategic ambition.

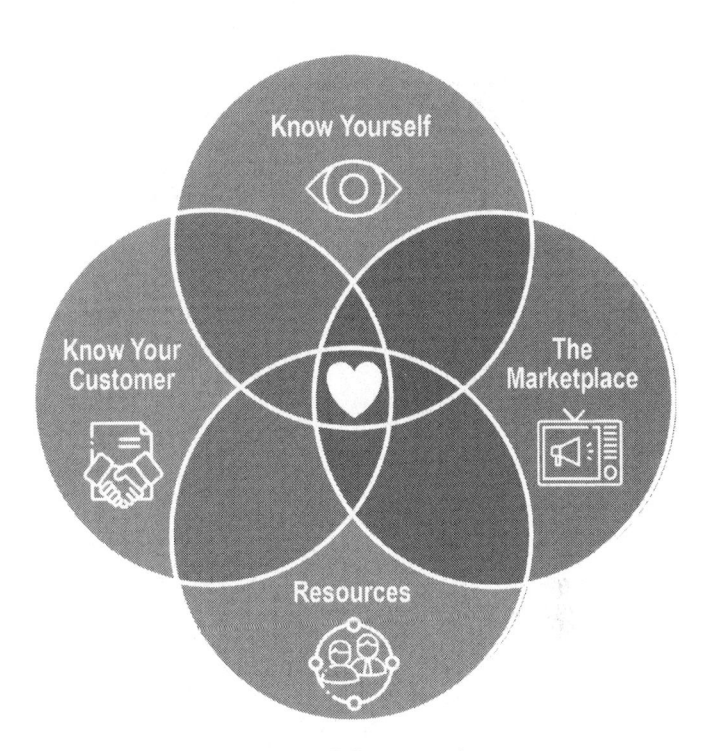

Figure 7.2. Micro-analysis.

Click to Transform

5. *Current Position*, Use your current web analytics to find your starting position.
   - Advocacy—slow but steady growth
   - Authority—the preserve of a few focused and dedicated innovators in any market sector
   - Attention—the default choice for those who need quick wins and have yet to truly innovate or disrupt
   - Prime—the largest, most well-known dominant player

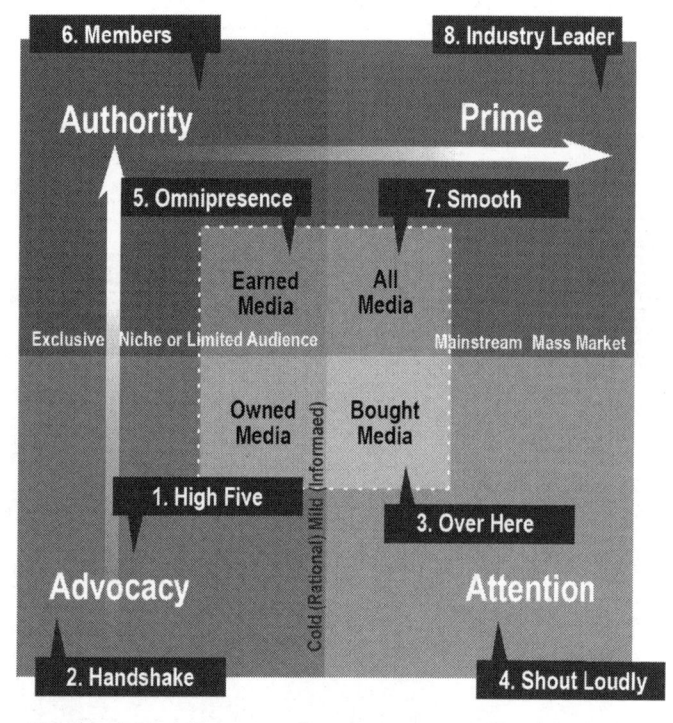

Figure 7.3. Macro-analysis using ionology strategy.

6. *Engine of growth.* Once you know where you are, you can make a play and move your market position.
7. *Tactics.* Create mile markers. Break them into projects. Break the projects into tasks.

Culture, People, and Alignment

From the digital business strategy must come the right story, focal points, and vision that change will bring. Clarify with every individual their understanding of the strategy and the role they play in creating the new vision. Explain how their input will be measured and validated. Give them the tools, training, and support they'll need in order to achieve their personal purpose and explain how they stand to progress in their careers by transforming with the new opportunity.

Establish the team, and allocate every individual one task each week that is strategic but not part of their normal job. Small, frequent, but important tasks allow the organization to change at a continuous, comfortable pace. Find the right internal advocates. Seek those with emotional commitment and an intellectual connection to the project.

Education

As well as personal betterment, a leader needs to understand the latest thinking beyond their own industry and appreciate how business models are evolving and changing because of digital disruption.

Individual team members should have a clear training agenda. The training requirements should focus on filling in gaps to allow people to best complete the challenges laid out in the digital business strategy.

Nurture Agile Innovation

Innovations = New Ideas Commercialized.

If new, 'innovative' ideas could be stacked on top of each other, they would fill any company warehouse within a week. The challenge isn't getting new ideas. It's getting ideas that overcome diagnosed business challenges extending from the digital business strategy.

Pursue Tactical Excellence

The quality of those tactical outputs depends very much on the quality of the leadership and their ability to articulate their strategy. With a clearly defined strategy and trained marketing and technical professionals, the delivery of superb tactical work is typically ensured.

UCLA Management Professor Richard Rumelt suggests in his highly acclaimed book *Good Strategy/Bad Strategy* that a good strategy should contain a strategy kernel. The "diagnosis" defines the nature of the challenge, the "guiding policy" is the overall approach that can be taken to overcome defined obstacles, and "coherent action" is a set of coordinated actions and resources which overcome the diagnosed challenge.

Tactical excellence is achieved when the diagnosis is accurate and the guiding policy is clearly articulated, creating coherent action. The mile markers are broken into projects and are handled by project managers. The projects are broken into tasks. The tasks become coherent action. An individual or team is allocated a task with guidance on why they are embarking on that task and what success looks like.

Summary

Digitizing the customer experience and business operations is the central focus within the Cognizant framework. In three months, this framework allowed me to build SourceConnecte from scratch. The digitization of the supply chain provided a safety net that we quickly deployed in the face of COVID-19. The national supply chain of PPE crumbled. SourceConnecte transformed it.

In Chapter 2, I briefly described how we digitized the customer experience by finding and connecting the right APIs. This was only possible by looking at the big picture, as described in Chapter 6. I identified the data sink, the processes they used, and how they handled data. The technology to make their jobs easier already existed. Trust Your Supplier, for example, organized and shared data through blockchain. I simply connected the technology in a way that offered a seamless, digitized customer experience.

Next, I looked for the opportunity to digitize the products and services offered. I pulled Inxeption into the web of APIs and integrated it accordingly. Inxeption now works with Trust Your Supplier and IBM Sterling, and not a single person was hired in the process.

After a three-month pilot program, SourceConnecte became fully operational. The timing was eerily convenient. COVID-19 broke the traditional supply chains of PPE. Manufacturers who earlier had no interest in PPE were now in the market. Companies who needed PPE couldn't find it anywhere. Here we were. Other supply chains had to spend time digitizing their processes to adapt to the "new normal." SourceConnecte never knew the "old normal," so we were already using contemporary innovations to set up systems that were ready for COVID-19. We were able to fix those supply chains and connect manufacturers with companies whose supply had dried up. None of this would have been possible without a full range of digitized products, customer experience, and operations.

In retrospect, the Cognizant framework aligns with our strategy of operating in a virtual space to maximize our market exploitation. Our model was built on the fact that supply chain operations were no longer bound by physical locations. But it is not the only framework capable of transforming an industry. Before you click to transform, take a step back. The individual pieces of your framework will come into place. Your job is to evaluate where your industry stands. You need to consider your desired results. Only then will you find the right framework and gather the technology that fits.

8

Embrace Transformation

From a business perspective, differentiating business processes and quality customer service are central to overall success. Business leaders must therefore clearly identify and measure how information technology contributes to the value of every key business process. They must also know how to most cost-effectively use IT when the task is merely the management of commodity operations.

Just focusing on infrastructure improvements may result in cost rationalization, but it can also obscure the impact and value of applications and business processes to the end customer. Quality of service is always an essential ingredient in evaluating the business effectiveness—the elements of which are infrastructure, resources, activities, and services that span the entire business lifecycle.

Business leaders must embrace digital transformation because the right blend of cloud, managed services, and traditional privately run datacenters will deliver the following:

- An ability to leverage economies of scale across the service ecosystem created by using multiple cloud service providers
- An understanding of business value that expands the traditional financial values of the total cost of ownership (TCO) and return on investment (ROI) by including customer value, seller provider value, broker value, market brand value, corporate value, and the technical value of any investment
- A wider view of technology's impact on a business through the acceptance of a business as a portfolio of business processes that demand the use of portfolio management techniques
- An understanding of why business processes and their associated IT investments should be classified as differentiating based on IT, differentiating not based on IT, or not differentiating at all.

When viewed from this lens, digital transformation delivers business and mission value by doing these:

- Accelerating speed to market
- Strengthening competitive positioning
- Boosting revenue growth
- Raising employee productivity
- Expanding the ability to acquire, engage, and retain customers

Success, however, requires these:

- Envisioning transformation as an economic and business process improvement revolution, not a technical one
- Relying on metric-driven goals and plans which are explicitly driven by the organization's goals
- Ensuring organizational goals are compatible with cloud business enablers
- Ensuring enablers support the overall business strategy and align with the available economic options for consuming cloud services

Summary

This book is based on the knowledge that I have been collecting for over forty years. Not all of this knowledge was available at the beginning of my career. I have personally witnessed the evolution of the World Wide Web, the Internet, and cloud computing. Before starting SourceConnecte, I started four other companies. During this time, I wasn't held back by a lack of information or slow-moving technology. I was only ever held back by fear.

This is the luckiest time to be alive. Technology has advanced so far that it is no longer relevant to building a company that disrupts an industry. Our main focus can now be shifted to the relationships, connections, and concepts that will shape the future. Connections and relationships are no longer limited by physical distance. Innovation can now exist on a global scale. It's possible to transform an entire industry without writing a single piece of code.

Transformation isn't just about creating something new—it's also about replacing how things are currently done. Stepping into the future means stepping out of the past. We are being propelled into the future at speeds unlike any age in history. Do not let any of this scare you. Fear is rooted in a lack of knowledge. You know what steps lead to transformational innovation. Identify data sources and sinks. Use operational metrics. Embrace a hybrid IT strategy that involves cloud computing. All these processes have been explained so that you can enter any industry and solve its problems.

All you have to do now is click to transform!

Acknowledgements

I'd like to acknowledge Melvin Greer, Scott Goessling, Ericcson, AT&T Business, Intel Corporation, (ISC)2, and Tulane University SoPA for helping me along my transformational journey.

About the Author

Kevin L. Jackson is a globally recognized thought leader, industry influencer, and the founder/author of the award-winning *Cloud Musings* blog. He has also been recognized as a "Top 5G Influencer" (Onalytica 2019, Radar 2020) and a "Top 50 Global Digital Transformation Thought Leader" (Thinkers 360 2019), and provides integrated social media services to AT&T, Intel, Broadcom, Ericsson, and other leading companies. Kevin's commercial experience includes positions such as the Vice President of J.P. Morgan Chase, Worldwide Sales Executive for IBM, and SAIC (Engility) Director for Cloud Solutions. He has served on teams that have supported digital transformation projects for the North Atlantic Treaty Organization (NATO) and the US Intelligence Community.

Kevin's formal education includes a MS in Computer Engineering from Naval Postgraduate School; an MA in National Security & Strategic Studies from Naval War College; and a BS in Aerospace Engineering from the United States Naval Academy. Internationally recognizable firms that have sponsored articles authored by him include Cisco, Microsoft, Citrix, and IBM. His books include *Practical Cloud Se-*

curity: A Cross Industry View (Taylor & Francis, 2016) and *Architecting Cloud Computing Solutions* (Packt, 2018). He also delivers online training through Tulane University, Germanna Community College, O'Reilly Media, LinkedIn Learning, and Pluralsight.

Kevin retired from the U.S. Navy in 1994, earning specialties in Space Systems Engineering, Carrier Onboard Delivery Logistics, and carrier-based Airborne Early Warning and Control. While active, he also served with the National Reconnaissance Office and Operational Support Office, providing tactical support to Navy and Marine Corps forces worldwide.

Kevin L. Jackson

GET YOUR FREE DOWNLOAD AT:

CLICK TO TRANSFORM

LEADING TRANSFORMATION

Kevin L. Jackson

www.leaderspress.com/leading-transformation